Unmanning

War Culture

Edited by Daniel Leonard Bernardi

Books in this series address the myriad ways in which warfare informs diverse cultural practices, as well as the way cultural practices–from cinema to social media–inform the practice of warfare. They illuminate the insights and limitations of critical theories that describe, explain and politicize the phenomena of war culture. Traversing both national and intellectual borders, authors from a wide range of fields and disciplines collectively examine the articulation of war, its everyday practices, and its impact on individuals and societies throughout modern history.

Unmanning

• •

How Humans, Machines, and Media Perform Drone Warfare

KATHERINE CHANDLER

Rutgers University Press

New Brunswick, Camden, and Newark, New Jersey, and London

Library of Congress Cataloging-in-Publication Data

Names: Chandler, Katherine, 1978– author.
Title: Unmanning : how humans, machines and media perform drone warfare /
    Katherine Chandler.
Other titles: How humans, machines and media perform drone warfare
Description: New Brunswick : Rutgers University Press, [2020] | Series: War culture |
    Includes bibliographical references and index.
Identifiers: LCCN 2019021155 | ISBN 9781978809741 (paperback) |
    ISBN 9781978809758 (cloth) | ISBN 9781978809765 (epub) |
    ISBN 9781978809789 (pdf) | ISBN 9781978809772 (mobi)
Subjects: LCSH: Drone aircraft—United States—History—20th century. | Drone aircraft—
    Case studies. | Human-machine systems—United States—History—20th century. |
    Uninhabited combat aerial vehicles—United States—History—20th century. |
    United States—Military policy.
Classification: LCC UG1242.D7 C435 2020 | DDC 358.4/183—dc23
LC record available at https://lccn.loc.gov/2019021155

A British Cataloging-in-Publication record for this book is available from the British Library.

♾ The paper used in this publication meets the requirements of the American National
Standard for Information Sciences—Permanence of Paper for Printed Library Materials,
ANSI Z39.48-1992.

www.rutgersuniversitypress.org

Manufactured in the United States of America

Contents

Unmanning

Introduction

•••••••••••••••••••••••

A "Different Lethality"

On February 6, 2002, American television news announced that Osama bin Laden may have been killed by a missile fired from a Predator drone two days earlier in southeastern Afghanistan. Confirmation was delayed, according to the reports, due to poor weather and the inaccessibility of the region.[1] A week after the attack, on February 11, an article in the *New York Times* described "troops of the 101st Airborne Division . . . at the grisly task of gathering evidence at the campsite where a missile-carrying American Predator drone fired at a small band of suspected members of Al Qaeda." While it downplayed earlier speculation "that a tall bearded man seen through the Predator's camera" was bin Laden, the article nonetheless emphasized that the drone weapon system had been fashioned to target the leader of Al Qaeda.[2] Pointing out that bin Laden had been captured by cameras on the unmanned platform previously, officials proposed that "arming the Predator seemed to provide the answer: Mr. bin Laden and his lieutenants could be attacked by the same surveillance drone that spotted them."[3] This statement figures the action of the terrorist target and the Predator drone as if one were the counterpart of the other. Uncertainty about who was killed by the unmanned aircraft, apparent after the arrival of ground troops, did not unsettle the potency of the connection between the two.[4] The Predator drone enmeshes aircraft, camera, and operator to propose an alignment between a visual field onscreen with the political act of defining an enemy as if determined by machinelike feedback and objective fact.

Secretary of Defense Donald Rumsfeld testified to Congress in support of the new technology just after news reports broke the story of the strike in

Afghanistan. He used the logic of integrated targeting and surveillance to support new spending: "If you have an unarmed Predator that's out there gathering intelligence information and you replace it with an armed Predator, that not only can gather intelligence information, but then can actually fire a Hellfire . . . you've got *different lethality*."[5] Grégoire Chamayou writes, at its base, the drone apparatus "is nothing less than an understanding of politics."[6] Yet, unmanning takes the guise of an inhuman response that evacuates the political. Rumsfeld's testimony proposes different lethality as a seamless field, linking information with targeting such that human action appears instead as a closed network between Predator and Hellfire technologies. This mandate is articulated in the context of the 2003 Department of Defense budget. Unmanning is premised on the undoing of human action as technological optimization; the basis of its politics is a disavowal of how connections between humans, machines, and media make the negative affect unmanning names.

The apparently mechanical integration of targeting and surveillance, promoted as a novelty at the outset of the war on terror, does not displace human for machine but, rather, confuses and denies one for the other. With the drone, what is *not human* is performed by human, machine, and media to minimize the politics at its basis. The multiple and changing ways the drone performs this politics is outlined by shifts in what the drone is over the twentieth century. In the 1930s, the U.S. military experimented with the drone as a radio-controlled target to be shot at, later integrating television to produce an assault weapon. At the end of World War II, the military halted efforts to build drone weapons that were based on the target, which were deemed misguided failures. The stakes of replacing humans with machines were not clear, nor was success inevitable but a contested struggle situated in global transformation. In the Cold War, unmanning reemerged as a platform for surveillance, integrating aerial photography as an alternative to piloted reconnaissance flights. Experimental aircraft built for these purposes were flown for thousands of missions in Vietnam before the program ended. Both the interwar and Vietnam-era phases of experimentation construct the drone through a disavowal of politics by technology. The idealized negation of the human sought in the early processes of unmanning is never achieved. These failures underscore the performative dimensions of unmanning and how the drone is animated by a political context. Experimental missions with the Predator similarly rely on an idealization of technology to justify the targeting of bin Laden and underplay the politics that drives its use.

During unmanned flights in 2000, a Predator aircraft captured hours of video footage of bin Laden at Tarnak Farms, outside Kandahar, Afghanistan. According to officials familiar with the video, the system transmitted "live pictures of a tall, white-robed man surrounded by a security detail" at the Al

Qaeda training camp.[7] The video circulated within the defense community and, reportedly, at a viewing with the secretary, chief of staff, and assistant vice chief of staff of the air force, the addition of a laser-guided Hellfire missile to the drone was proposed.[8] In February 2001, the U.S. military tested the Predator aircraft and Hellfire missile system at Creech Air Force Base in Indian Springs, Nevada. In the aftermath of the September 11, 2001, attacks against the United States, industry and military advocates for the Predator seized on footage of bin Laden to catalyze preexisting development of the unmanned program.[9] In this way, some military and government officials had anticipated bin Laden's death by drone before it was announced on the news on February 6, 2002, even as they assembled the Predator system through human (viewer/operator), machine (aircraft/Hellfire missile), and media (camera/video).

Thus, the drone's so-called technicity and "new" paradigmatic role in the war on terror emerged before war was declared. Rather than a mere response to the battlefield, a future scene of attack was created in the Predator's development. Rumsfeld's testimony describes the Predator as a closed loop defined by intelligence and targeting, decoupled from its interpretative context. A political scene—attacking bin Laden—is recounted as technological advance. Unmanning performs a dual separation of technology and politics. First, unmanning collapses the operator, aircraft, and camera into a single technological unit. Second, unmanning decontextualizes the drone as a novelty, distinct from the numerous iterations that presage contemporary unmanned aircraft. Whether the strike goes as intended or not, the target is imagined not just through a picture but through the integration of human, machine, and media that makes the drone itself.

On February 12, 2002, the Pentagon issued another official statement about the strikes in Afghanistan, insisting that the Hellfire missiles fired by the Predator system a week before had hit their targets. "Those killed were 'not innocents,'" said Rear Adm. John D. Stufflebeem, deputy director of operations for the Joint Chiefs of Staff in a press conference."[10] He explained, "I base that on the *facts* that [the 101st Airborne] . . . did some exploration in the surrounding area, to include some caves, a nearby village, and talking to locals. So I think that that sort of puts us in a comfort zone. These were *not innocents*."[11] This response came after news articles (first reported on the front page of the *Washington Post* on February 10, 2002) indicated that the persons killed in the strike were civilians.[12] Instead of bin Laden or his associates, the "victims might have been scrap metal dealers or smugglers searching for weapons abandoned by Al Qaeda and Taliban troops several weeks [earlier]."[13] The debate between reports by the government and the accounts of the strike published by major newspapers after bin Laden's supposed death contradict the purportedly integrated field of targeting and surveillance as the Predator's basis. Rather, the battlefield is an interpretative context for the U.S. government, which establishes the facts of the case. The official account contrasts markedly with that of persons in Afghanistan.

A follow-up investigation published on February 17, 2002, in the *New York Times* further eroded the government's claims. John Burns writes, "A visit this week to the site of the missile strike, and to nearby villages, established that the men killed were Daraz Khan ... about 31, from the village of Lalazha, and two others, Jehangir Khan, about 28, and Mir Ahmed, about 30, from the village of Patalan." He remarks, "On a trip into the mountains, just about every Afghan encountered along the way—goatherders, brushwood collectors, militiamen and farmers—readily identified the three victims, and their villages." Countering the official assertion that the men were affiliated with Al Qaeda, Burns notes, "The idea [was] dismissed with bitter mockery by the dead men's family, neighbors, and local militiamen, who say the victims were poor villagers with no history or interest in militant Islamic politics." The cave complex where the men were killed was the site of a former guerilla camp, which had been bombed by U.S. forces a month earlier. Metal remnants collected from the abandoned site could be sold across the border in Pakistan for forty or fifty cents a load. The article notes a final tragic detail: Daraz Khan was known locally as "Tall Man," which perhaps accounted for the mistaken assumption that he was bin Laden, also known for his height.[14]

In the context of the war on terror, bin Laden's earlier capture by surveillance cameras in 2000 makes possible the claim that he was killed in 2002. The U.S. government proposed the strike's basis was the "fact" that Daraz Khan, Jehangir Kan, and Mir Ahmed were persons "not innocent," which overwrote the account given by their families and community. The Predator drone strike not only killed, but also made a political determination as to who was and could be a target. That bin Laden might have been pictured by the drone's cameras justified the deaths; objectivity was asserted by the technological possibility backed by governmental authority. Misperception, as much as identification, is at play in naming the threat "terrorist." Reactions to the targeted strike in 2002 resonate with contemporary concerns about targeted killings, namely, that all military-aged men pictured by the drone's cameras can become targets. Speaking on February 12, Victoria Clarke, Rumsfeld's press secretary, contended, "We're convinced it was an appropriate target, based on the observation, based on the information that it was an appropriate target."[15] Government spokespersons further attempted to counteract the villagers' narrative by suggesting that local Afghans were unreliable, explaining that their claims might be an attempt to gain compensation from the United States. The rubric of drone killing, here, maps onto decades of U.S. military interventions discounting local claims and knowledge in a mirror of histories of colonialism.

Rumsfeld posited that the effectiveness of the drone system was based on the seamless coordination of intelligence and targeting. Derek Gregory examines this claim in his seminal article on drone warfare a decade later, identifying it as "a view to a kill." He writes of the integration of surveillance and targeting, "these visibilities are necessarily conditional—spaces of constructed visibility are also always spaces of constructed invisibility—because they are not technical but rather techno-*cultural* accomplishments."[16] His argument emphasizes how colonial histories layer into drone technologies and their hunter-killer function. By focusing on the view, however, an image seems to underlie both the practices of watching and targeting. Yet the onscreen picture deteriorates when counterposed with the account given by Khan's relatives and the fact that bin Laden was not killed in 2002. What organizes the strike is not "a view" but "the drone" and the network of humans, machines, and media that makes it. The context is interpreted not through information but by the "different lethality" provided by the Predator. The substantial link between the scenes captured by the Predator's surveillance cameras in Afghanistan in 2000 and the Hellfire missile strike in 2002 is not the presence of a terrorist leader and his picture onscreen. Rather, the commonality is the drone system and the actions taken by the U.S. government. This book examines how enemy and territory are made by drone parts that contest the apparently neutral and objective view proposed for the Predator.

Burns's article on the aftermath of the strike quotes Amir Khan, a local militia commander in Kandahar, Afghanistan, who states, "This is theater, what the Americans have done here, just theater."[17] Khan's point stresses that the apparent technicity of targeted killing is performed. The first reported strike by a Predator illustrates the tragedy of targeted killing, but it also suggests a script. The overlay between surveillance and targeting enables the U.S. government and military to claim an all-the-more total view of the battlefield, even when the picture does not correspond with the ground below. In hindsight, the Predator's images appear far from complete, questioning rather than reinforcing the totalizing view claimed for "different lethality." The previous account outlines how drone aircraft perform a scene of war they are simultaneously imagined to counter, while neither corresponds with conditions on the ground. The historical case studies in the book show how this function transforms and develops over the twentieth century. This book's aim is not just to point to the technocultural assumptions that are built into the view transmitted by the drone but also to assert how the drone itself is fashioned by technoculture and its ties to changing relations between the United States and the globe.

Despite its prominence in newspaper reports, the strike in February 2002 received relatively little follow-up. Most writing on drone aircraft dates the first targeted killing to a strike several months later in Yemen, a killing that took place outside the declared conflict in Afghanistan.[18] On November 3, 2002, news media announced that a Predator strike had killed a suspected Al Qaeda member, Qaed Salim Sinan al-Herethi, along with five associates in his car. In this case, the response was markedly different, as no official comments were made. The *New York Times* noted, "A C.I.A. spokesman refused to comment today. Nor would White House officials confirm the Predator strike in Yemen or say whether President Bush had personally authorized it."[19] In the attack against al-Herethi, forensic evidence supersedes the Pentagon's previous insistence the unmanned system killed "appropriate targets." The death of a target implicitly justifies the strike, and assurances of the strike's legality are given off the record.[20] Here, the closed loop between surveillance and targeting appears to function autonomously, while the role of political decision-makers is classified. It is worth recalling that in the strike against al-Herethi five other persons were killed, including U.S. citizen Ahmed Hijazi. Investigations by the American Civil Liberties Union and the United Nations contest the official silence and apparent legality.[21]

Today, drone warfare has become a significant framework for counterterrorism. The legitimacy of the program is premised on targeting suspected militants. Yet, little information has been made public about the thousands of persons killed since 2002, and the program is marked by repeated denials of illegitimacy ("those killed were 'not innocent'"). The paucity of detail and secrecy has been exacerbated under the presidency of Donald Trump.[22] Drone serves as shorthand for the integration of operator, camera, and aircraft, as if they were a single, machinelike unit, while its target is imagined as a singular terrorist. Yet, neither of these images hold up. This book turns to a genealogy of drone aircraft and the secret projects from eight decades of experimentation between 1936 and 1992 to interrupt these images of contemporary targeted killing. Through the ways drone aircraft have been made, broken down, and remade, I reconsider unmanning through the concept's emergence in the twentieth century and its ties to a politics disavowed. Unmanning is organized by the drone's parts and their contradictions. To study and make explicit these paradoxes works against official denials and silences, counteracting unmanning's purported undoing of human action. Drone warfare is haunted by violence that exceeds the current war on terror, made possible through confusions between humans and machines.

Questioning Technology and Politics

The question of how the drone is made draws on the framework of science and technology studies (STS), which emphasizes the irreducibility of humans and nonhumans and questions categorical distinctions between human practices, scientific facts, and technological objects. Donna Haraway's "A Cyborg Manifesto," for example, positions humans as cyborgs, entanglements of persons and technologies, while Bruno Latour's *We Have Never Been Modern* rewrites the modern constitution to unsettle a divide between the production

of social values and scientific fact.[23] Rather, Latour argues, collectives are organized through processes that integrate human and nonhuman. Networked
relations are the basis of STS. These theories lead to an articulation of the
co-production of technologies and social order. In *States of Knowledge*, Sheila
Jasanoff observes, "Co-production is shorthand for the proposition that the
ways we know and represent the world (both nature and society) are insepa-
rable from the ways in which we choose to live in it." Science and technology,
she explains, "[embed and are] embedded in social practices, identities, norms,
conventions, discourses, instruments and institutions." [24] Technical and political practices are, therefore, inextricable from one another.

Unmanning is indebted to the networked assemblages outlined by STS
theories. Pointedly, there is no unified "drone"; rather, it is made by human,
machine, and media parts that take distinct form in these case studies. The
networked system might be an instance of Haraway's cyborg, tying together
human and machine in a lethal form created by military industry. The study
of unmanning, however, attends to the ways drone networks not only come
together but also *fail to cohere* and the attendant politics that emerge through
these relations. Charis Thompson observes in her study of gender and technology, "Attempting to elucidate some of the specific choreography that
enables ontologically different kinds of things to come together has inevita-
bly led me to explore the ontological separations between things . . . and to
examine the reductions of one kind of thing to another." [25] *Unmanning* shifts
attention from the coproduction of technology and politics to also examine the negations and confusions that develop between the drone's human,
machine, and media parts. The technopolitics made by the drone are tied to
confusions and ambiguities that are part of the drone's ontology, not only
establishing what is unmanned but also proposing an ideal of what is human
as unmanning's counterpart.

I expand on the idea of ontological choreography by showing how unmanning is established not just through a synthesis of human and nonhuman but
through an undoing that also inheres within drone experiments and their
political consequences. How does the drone produce what it is *not*, and how
does this contradiction play out in its wartime uses? These questions return to
divisions between humans, technologies, and politics long questioned by science and technology studies, not to reify their difference but to consider how
their relations are also constituted through disavowal. Rather than take up
just how humans and machines come together to produce declarative forms
of knowing and representing, I use the concept of unmanning to examine how
humans and machines can also be used to deny and misrepresent. This move
draws on feminist approaches to science and technology studies, which analyze affective dimensions of technoscience and their social mapping of power.
Disavowal also draws from psychoanalytic approaches within poststructuralist

and feminist theory. Disavowal decenters knowledge claims made by techno-science and politics through situated experiences that do not cohere and, thus, enact contradiction. It works against the collapse of aircraft, camera, and operator and the supposedly seamless field of targeting and surveillance that the contemporary drone system organizes.

In psychoanalysis, disavowal is a specific mode of defense that consists of the subject's refusal to recognize a traumatic perception. Freud uses the term to name the male subject's refusal to see an absent penis in females.[26] The concept of unmanning, however, troubles and reproduces "man" through the opposition between human and machine. It is both incidental and cogent that unmanning names a gendered position as its opposite. While the difference between male and female is itself disavowed in Freud's theory, unmanning avows a distinction between human and machine to reiterate "man." The exchange between human and machine relies on contradiction, not identification, and the drone is an assemblage of parts, not a subject. The prefix *un-* is a negation; *un*manned names the opposite of manned. Unmanning is, thus, structured as a denial, both negating and acknowledging what it is not. It emphasizes the political use of the term rather than its psychoanalytic specificity. "Man" is what is undone by the parts of the drone and is, as such, all the more insistently its backdrop. This is not to claim the primacy of human action or the "social," claims long troubled by science and technology studies. Rather, it is to think about how human negation is made a counterpart of technoscience and what it means for human actions—and politics—to occur in this supposed absence.

The disavowal articulated by unmanning ties to more recent work by feminist queer theorists who complicate Freud's theories. Again, it is significant that the drone is an assemblage, not a subject, that nonetheless is posited at once as opposite to and aligned with "man." In their writing on cross-identification, Judith Butler and Biddy Martin observe that "every identification presumes a crossing of sorts, a movement toward some other site with which or by which an identification is said to take place. But it is because this 'crossing' is understood that we underscore it through redundancy here. 'Crossing' may be conceived, on the one hand, as an appropriation, assimilation, or even a territorialization of another site or position, or it can be understood as a movement beyond the stasis attributed to 'positions' located on a closed map of social power."[27] The back and forth between what is human and what is machine, what is man and what is unmanned, organizes power. While unmanning proposes a technopolitical order, disavowal builds a negative field of denial that reasserts "man" through his supposed undoing.[28]

Human, machine, and media parts that make drone aircraft are contingent, unpredictable, and spectral, even as they propose a rational and calculated framework for war. With these contradictions, unmanning does not only disavow "man," although this is its most literal claim. As a map of power, the disavowal named by unmanning is also tied to the trauma of war and its violence. The term is significant not only for its gendered connotations but also for the trauma it denies. Technology is not only "unmanned" in the case of the drone but also produced as the counterpart to threat, as in the presumed relationship between bin Laden and the Predator drone discussed earlier. The development of drone aircraft in the twentieth-century United States proposes a kind of war that fits with a set of technoscientific approaches, aiming to normalize aerial attack, nuclear missiles, and automation as means of global control and national defense. Ian Shaw describes the drone as part of a technological enframing, noting how its terms equivocate killing with a machine-like response and survival. He writes, "Targeted killings are quickly becoming a 'post-political' background and a noise few listen to."[29] Attending to the negated human figure in unmanning troubles these automatic operations. In contradistinction to Shaw's claim, politics continues in denial. Although unmanning claims to be not "man" and insists on a field of war defined by objective fact, the contradictions that inhere in the term show the persistence of human action.

The organization of power through disavowal, in this way, extends technoscience to necropolitics. Achille Mbembe's reformulation of politics as a rendering of death draws together theories of biopower, sovereignty, and the state of exception. Mbembe asks, "What place is given to life, death, and the human body? How are they inscribed in the order of power?"[30] Through his theorization of necropolitics, Mbembe attends to the shadows of the modern state and political subject—the plantation and colonies. By observing how the sovereign subject is set against the denial of humanity under slavery and colonialism, Mbembe argues that death haunts the modern political order. In the post-colony, Mbembe sees death's reemergence of slavery and ties necropolitics to aerial policing and urban siege. The machinelike framework of the drone is doubled to establish threat as inhuman, even as denial never fully erases humanity. Aircraft, operator, and camera render a person killable. Targeted killing interweaves politics, technoscience, and trauma, even as the use of the system is described as a technological calculus that weighs threat and protection. This denial builds on longstanding practices of violence that rely on the divide between human and nonhuman, which resonates in unmanning. Frantz Fanon observes, "Sometimes this Manichaeanism reaches its logical conclusion and dehumanizes the colonized subject. In plain talk, he is reduced to the state of an animal."[31] In the twentieth century, the drone is a case study in how machines come to form part of this dualism, while vestiges of animalization remain in the insect, bird, and hunting analogies that name the aircraft.

The negated field of politics named by unmanning maps human and non-human against the backdrop of gender and colonialism. The shadows cast by disavowal make and unmake "man" and "enemy." These figures haunt its tech-noscientific framework. They recall what Avery Gordon in *Ghostly Matters* describes as "that special instance of the merging of the visible and invisible, the dead and living, the past and the present. . . . It is a case of the difference it makes to start with the marginal, with what we normally exclude or banish, or, more commonly, with what we never even notice."[32] The violence outlined by unmanning exceeds the technopolitics that drone aircraft produce. Working from the failures of drone projects in the twentieth century—all programs that had little strategic significance in war—my analysis considers how they nonetheless provide a specter of contemporary violence and challenge the dehumanization often used to account for the rise of drone aircraft. Rather, human action has animated drone flight. The result, as Katherine Kindervater writes, "is not bureaucratic death-dealing with an increasingly automated and unaccountable decision-making devoid of human thought, but instead the culmination—or clear expression—of the violence of Western thought itself."[33] As an assemblage of human, machine, and media parts, it is not just the coherence of violence that inheres in the drone but also a map of power that is defined by disavowal and its haunting.

Drone Mediations

Media extends from the human and machine parts of the drone to a mediated context. In this way, I add to the framework of STS through media theory. Writing in 1964, Marshall McLuhan theorized media as human extension. He observed, "Long accustomed to splitting and dividing all things as a means of control, it is sometimes a bit of a shock to be reminded that, in operational and practical fact, the medium is the message. This is merely to say that the personal and social consequences of any medium—that is, of any extension of ourselves—result from the new scale that is introduced into our affairs by each extension of ourselves, or by any new technology."[34] Media situates drone aircraft in radio, television, programmed control, photography, and video. Human, machine, and media parts do not just form the drone but scale out to everyday life and the shaping of affective experiences in the twentieth and twenty-first centuries. Whereas the opposition put forward by unmanning bifurcates human and machine, media theory shows how this tension inter-twines militarization with everyday life in the United States.

Mediation extends beyond a mere mode of communication between human and nonhuman, as it is set up in drone experiments to also tie to a theater of war that it produces and is produced by. Friedrich Kittler writes, "Media determine our situation, which—in spite or because of it—deserves a description."[35] Purposefully, his argument locates media within its military uses and the determinism it proposes. Military and government accounts of drone aircraft are predicated on the alignment between seeing, knowing, and sovereignty as if its mediations provided a god's-eye view and an objectively given reality. Yet, this alignment does not always fit, and as Kittler notes, media stories might rather be "a ghostly image of our present as future."[36] In this way, I emphasize not the determinism of media but its situatedness. Haraway's essay on situated knowledges critiques the god's-eye view of military and science to ask instead how vision is made. She writes, "Struggles over what will count as rational accounts of the world are struggles over *how* to see."[37] Media theories, distinct from science, military, and government accounts of seeing, have long grappled with mediation, its limits, and its ties to embodiment. Drone mediations are, then, fraught with variability, contradiction, and error. Rather than producing a singular aerial view from nowhere, these experiments are tied to specific scenes of war and their undoing.

The case studies in this book outline how the myth of unmanning fails over the course of the twentieth century. They complicate Jeremy Packer and Josh Reeves's claim that "with the rise of the drone, Earth observation media are . . . now being integrated with automated attack capacities so that they can most efficiently track and eliminate the enemy no matter where it might hide."[38] Against this ideal, operators and other analysts of unmanned aircraft emphasize the mediated limits of drone aircraft.[39] The drone and the field of war to which it is tied are contradictory and error-prone and, yet, nonetheless mediate and organize politics. Lisa Parks observes, "A critique of vertical mediation involves explicating the kinds of capacities and power relations the aerial and orbital machines are used to enact or mobilize, while also remaining attentive to the limitations or failures of these technologies."[40] How drones have been imagined, made, and remade over the course of the twentieth century shows how drone mediations are built on breakdown and reveals their affective and subjective limits. It is through erasure and disavowal that the relations made by the drone are refashioned as all-seeing power.

As Caren Kaplan has noted in her analysis of the "myth of precision," aerial bombardment in World War II and geoinformation systems in the Cold War were ineffective. Both the Norden bombsight in World War II and Patriot missile attacks in Iraq are at odds with their mythos. Instead, Kaplan emphasizes how the mediation of war fits with everyday experiences. While the systems did not achieve their purported aim of precision on the battlefield, they created the context in which " 'precision' [had] entered the emotional field of subjectivity as the military-industrial complex [grew] to encompass more fully the culture of industries of media and entertainment."[41] Unmanning is achieved by mediation and its ambiguity; sight does not exactly equate with power nor does the drone immediately align with global control. Rather, drone aircraft create a context that describes the replacement of human for machine action, even as this disavowal breaks down and the mediated field of the drone remains contested and contradictory. This context not only plays out in war but also is embedded in media and everyday life.

A Genealogy of Drone Failure

In 1988, Congress commissioned a report classifying all current and historical unmanned aerial vehicles (UAVs). It included more than a hundred models, produced in the United States and abroad.[42] The report details various experiments with teleautomation, aerial torpedoes, drones, cruise missiles, and unmanned aircraft proposed since the early days of flight. It suggests a progression of technologies, transformed through increasingly sophisticated communication and computational systems, leading to contemporary unmanned aircraft. To describe the development of contemporary targeted killing, histories of drone aircraft like the report piece together technical details from such a patchwork of experimental aircraft.[43] Earlier accounts of these histories would not have necessarily positioned them in relation to the military strikes carried out today, as the term *drone* named target planes, prototype guided missiles, and other miscellany. The government report itself is based on a decision by Congress in 1987 that defined drones as reusable airframes, attempting to distinguish unmanned air vehicles from other pilotless, airborne systems such as cruise missiles.

My analysis of unmanning in these chapters does not outline systemic advance; instead, I emphasize the drone's performative dimension, complicating and undoing definitions of the term. Chapter 1 begins with "drone," which is not the first unmanned system built in the United States—there are many before—but is, rather, the first to use the term. The term names a U.S. Navy project to build a radio-control aircraft. The early "drone" was developed by the U.S. Navy as a target to be shot at, distinct from the contemporary Predator and with no camera or missile onboard, at least initially. For personnel trained in the 2000s to fly unmanned aircraft, the dissimilarity was fundamental. Predator pilots wore patches with the motto "We're not drones . . . We shoot back"[44] to distinguish the systems they operated from drones used for air defense training. This disjuncture highlights how the predecessors in the history of unmanning do not culminate in the drone system used in the war in terror. Rather, they fit within the concept of genealogy theorized by Michel Foucault, who notes, "What is found at the historical beginning of things is not the inviolable identity of their origin; it is the dissension of other things. It is disparity."[45] Discrepancies stress the ways that politics are enmeshed with drone systems and produce what come to be seen as the structuring qualities of aerial war, including the enemy target and the aerial view. The human, machine, and media parts of the drone are an "unstable assemblage of faults, fissures, and heterogeneous layers."[46] A genealogical approach addressing these paradoxical relations warrants studying what are typically presented as problems in the history of unmanned aircraft, namely, inconsistent terminology and a miscellany of air vehicles.

Early unmanned aircraft undo the commonly held idea that drone and terrorist are a reaction to one another and instead tie the drone to multiple enactments of aerial domination carried out by the United States over the course of the twentieth century. These cases are limited to the particular context of America in the twentieth and twenty-first centuries. Drones are animated variously as an insectlike target, a Japanese kamikaze, a cybernetic system, and a birdlike reconnaissance platform, making strange a system made by American military, industry, and science. Unmanned aircraft thus establish a context for the disavowal of politics and imagine the replacement of human for machine action. My research draws on five cases from this history to identify two key periods in the development of drone aircraft, defined by their proposed uses (which influence systems today): between 1936 and 1944, drones were training targets and weapon systems, and from 1953 to 1973, drones emerged as technologies for unmanned reconnaissance. Both weaponized and reconnaissance drones were dismissed as unsuccessful technological innovations during these periods, even though their uses presage today's unmanned combat air vehicles. The final period encompasses the development of the Israeli drones that progenerate contemporary systems, through weapons sales between the United States and Israel and their use in the First Lebanon War, often described as unmanning's first "success."

The first two chapters examine connections and ambivalences between drones as targets and as systems to target the enemy. They rely on the Collected Records of Delmar Fahrney, over one hundred boxes of materials gathered by U.S. Navy Rear Adm. Delmar Fahrney.[47] I use the archive to examine the navy projects Fahrney led, beginning in 1936. The collection was classified until 1995. Chapter 1, "DRONE," begins with the coining of the military term by Fahrney and a coworker at the Navy Research Laboratory in 1936. The first trials reveal how the drone both integrates and makes strange the drone's human, machine, and media parts. Crashes figure prominently in the early experiments, which disturbs the ideal of remote control captured by naming the system after a mindless, single-purpose insect. These early tests illustrate the changing strategies of airpower simulated by the radio-controlled planes. While most remotely piloted aircraft used during World War II were targets to train anti-aircraft gunners, U.S. Navy researchers, at the behest of Radio Corporation of America (RCA), also proposed that the system could be modified for use as a weapon. Chapter 2, "American Kamikaze," examines these projects, emphasizing the significance of RCA's television system. Technical discussions and military reports indicate how the assault drone was likened to racialized suicide bombings, as does the title of James Hall's memoir about the top-secret assault drone unit, *American Kamikaze*. While television-guided drones were correlated with enemy tactics, they were also promoted as exemplary of American ingenuity and moral superiority, using technology to eliminate human risk. Despite this positioning, the program was deemed a failure and ended in 1944.

Chapters 3 and 4 study how drone reconnaissance made oppositions between manned and unmanned flight during the Cold War, even as the systems entangled human and machine actions. This part of the book draws on documents from the Ryan Aeronautical Collection, originally established and maintained by William Wagner, corporate historian and public relations manager of Ryan Aeronautical.[48] A portion of this corporate collection is devoted to drone aircraft, and it draws on materials gathered for two books about the Ryan Aeronautical reconnaissance drone: *Lightning Bugs and Other Reconnaissance Drones* and *Fireflies and Other UAVs*, both written or cowritten by Wagner. The materials in the archive include lengthy transcribed interviews with the personnel involved in the projects, mostly conducted in 1971. Due to security restrictions, *Lightning Bugs* was not cleared for publication until 1982, and *Fireflies* was published in 1992. Additional documents include technical reports, company briefings, and promotional materials, as well as photographs and film footage. Declassified air force documents, including the "Buffalo Hunter" report and a history of the use of drones in Vietnam between 1970 and 1973, are also analyzed.

Chapter 3, "Unmanning," considers the use of drone aircraft as reconnaissance systems, linking machine autonomy with an aerial picture of national protection undoing the role of human action. Described as "a bee with an electronic brain," the drone is correlated with a cybernetic system to suggest that it operates on its own. Inputs and outputs that guide Cold War–era drones complicate a monadic cybernetic model. Rather, the human and machine actions that guide the drone are discursively performed as autonomous flight. After Francis Gary Powers's U-2 reconnaissance flight was shot down over the Soviet Union in 1960, unmanned surveillance was positioned as minimizing political risk. Because no pilot was onboard, the "machine" system was promoted as a technological alternative to politics. Chapter 4, "Buffalo Hunter," examines reconnaissance drones developed between 1960 and 1973, which were tested in the United States and flown in Southeast Asia. The chapters examine how national security is tied to reconnaissance, both as a response to threats to the United States and as the basis for international interventions in Vietnam. The aircraft are imbricated with postcolonialism and their pictures tied to erasures of people and territory below.

In 1971, Ryan Aeronautical unmanned aircraft were sold to Israel, and the Israeli military used the systems for surveillance during the Yom Kippur War in 1973. Chapter 5, "Pioneer," examines innovations in Israel that led to the development of a prototype of today's Predator drone. It studies the use of real-time video transmission from a drone platform deployed in the Bekaa Valley air battle in 1982. This chapter relies on documents from the Israeli Defense Forces Archive, as well as reports and articles for the U.S. military and press about the battle. These materials are read against newspaper reporting of the 1980s, which emphasize the excesses of unmanned aircraft programs, including overexpenditure and corruption. The chapter uses the rubric of corruption to counter the all-seeing eye of the drone, showing how what is seen through its human, machine, and media parts is limited and error-prone. Unmanning sets up a disavowal between what is human and what is not to establish conditions for contemporary targeted killing and overwrite a genealogy built on failures.

1

DRONE

•••••••••••••••••••••••

In the decade before World War II, the U.S. Navy launched a classified project known by the codename Drone. It took its inspiration from earlier attempts at unmanned aircraft that were discontinued by the military as failures. The project joined a multitude of prior, worldwide experiments in unmanned flight, which extended from aerostat flight in the eighteenth century and military experiments with air balloons in the nineteenth to aerial torpedoes built in World War I.

Project Drone, however, marks a discursive beginning in the history of pilotless aircraft: it is at this moment that the term *drone* begins to acquire the entangled layers of connotation that will lead to today's Predator drones, used in the United States's so-called war on terror. The Americans did not invent the secret name out of whole cloth; instead, they repurposed the term from a competing project begun earlier by the British. William Standley, the U.S. chief of naval operations, advocated for the experimental Drone project after attending the Second Naval Conference in London in 1935, where he witnessed anti-aircraft training with the Queen Bee, a remote-controlled plane meant to test ship defenses against aerial attack. In a memorandum he circulated to the navy on March 23, 1936, Standley wrote, "An urgent need in the fleet exists for radio-controlled aircraft for use as aerial targets."[1]

Delmar Fahrney, the officer-in-charge of Project Drone, recounted the conversation that led to the codename in a history of pilotless planes and guided weapons he wrote more than two decades later. "It was brought out that the English had dubbed their project the 'Queen Bee' and following this phraseology, a number of insect names were reviewed. It was decided that the word DRONE best fitted the situation in which a released target plane found itself engaged; and the terminology was easy to handle. Without further ado the name was used in all discussions oral and written and the term persists to this day."[2] The final declaration, "the term persists to this day," resonates beyond Fahrney's intended reader in the late 1950s. Indeed, "to this day," uses of the term *drone* proliferate and the insectoid analogy persists, describing military weapons, hobby planes used by enthusiasts, and an emerging commercial sector for surveillance, filming, and delivery. A difference inheres in the commonly understood purpose of the drone as offensive or, in its most benign form, watchful. "The situation" Fahrney describes posits the drone as a remote-controlled training target. While this use of the drone continues today, it has become less ascendant in the American mind. Neither Project Drone nor Queen Bee purported to be an offensive technology. Instead, the drone was proposed to simulate and study the conditions of a new field of battle: aerial combat. Project drone, in the interwar period, was an aircraft built to be shot down.

This chapter examines how the drone organizes a theater of war by suggesting that aerial offensive attack and defensive anti-aircraft response are a mechanical exchange. Theater of war is a concept that draws on Joseph Masco's study of "American self-fashioning through technoscience and threat projection."[3] During the interwar years, air power and its theater of war take shape in public trials of aerial bombardment and secret experiments with drone aircraft. These dual aspects produce a scene of battling machines and political decision-making as technoscientific observation. Yet, analyses of air power in the 1920s and archival documents from the target drone trials demonstrate that these scenes are far from mechanical. Instead, they are messy, human, and above all formed through and by discourse. In the interwar period, the target drone is shaped by anthropomorphism, theories of air power, divisions within the U.S. Navy, and the military's lack of prior experience with aerial war and simulated aerial bombardments. The theater of war created in anti-aircraft trainings was at once prescient of World War II, even as it utterly failed to prepare the U.S. Navy for air attack.

The memorandum in response to Standley that circulated in the Navy Bureau of Aeronautics highlights the drone's role in creating a theater of war. It explains that a remote-controlled aerial target was necessary for the following reasons: "1. Definite data must be obtained as to the effectiveness of present and projected anti-aircraft equipment before any further marked improvement can be reasonably expected. 2. Training of personnel assigned to anti-aircraft activities must be carried out *under conditions more closely simulating action conditions* than exist at present if maximum proficiency is to be obtained."[4] The rationale for the target drone in the memorandum highlights how remote-controlled aircraft are conceived as both defining and mimicking air power. Even as a training target to be shot at (and without a camera onboard), the drone informs the navy's emerging strategy of how to counter and participate in aerial bombardment.

In this way, the program is not just about creating a target or a targeting system when the platform is later transformed into an assault weapon—rather, it performs aerial battle from its inception.

The blurred lines between simulation, defensive response, and offensive attack are not unique to air power. Samuel Weber notes that the term "target" contains this conflation in its etymology. Its earliest usage in the 1400s meant "shield," which would have been used as a defensive technology in battle. Later, a target was used to practice archery, coming to primarily signify a point of aim in the middle of the eighteenth century. Weber draws out this implicit tension between defense and offense before turning to a *Washington Post* article describing the capture of Saddam Hussein and his advisers, " 'Target of Opportunity' Seized." He then ties the verb "to target" to a specific kind of knowledge: "The enemy would have to be *identified* and *localized*, *named* and *depicted*, in order to be made into an accessible target, susceptible to destruction." This kind of knowledge is not new to warfare. Yet, as Weber notes, aerial combat brought the "mobility, indeterminate structure, and unpredictability" of targeting to the fore.[5] While Weber points to Hussein's capture and the war on terror as the historical watershed for this new kind of indeterminate knowledge, the interwar drone's use as a target shows that these shifts unfolded decades before, in experimental efforts to build a remote-control plane in a theater built not only through identification and mobility but also through simulation.

The memorandum I quote above insists that the drone provide "definite data" in service of "maximum proficiency." As its progress in the interwar period shows, it does no such thing: the data remains murky; the proficiency, weak. In the early drone trials, contradiction and indeterminacy plagued the relationship between the drone-as-target and the navy's anti-aircraft targeting, while the "definite data" provided by the trials remained controversial and subject to interpretation within the naval hierarchy. The ambition to mimic "action conditions" via the target drone does not so much respond to those conditions as it imagines and creates an as-yet *undefined* context of aerial war. The drone—itself a patchy network of moving discursive parts drawn from human, machine, and media—thus shapes the context it is supposed to mimic, a theater of air war.

Inventing Air Power: From the *Ostfriesland* to Project Drone

In the 1930s, the destruction that would be wrought through aerial bombardment in World War II was not a foregone conclusion. Though the possibility of devastation from the air was widely imagined, military commanders had yet to reckon with the conditions of broad-scale aerial bombardment.[6] In World War I, Roger Ehlers points out that "senior officers viewed bombers as extensions of field artillery rather than independent bombing platforms."[7]

Airplanes primarily engaged in reconnaissance and tactical missions, using planes as spotters for artillery and troops on the ground. Aircraft flown by the U.S. Navy received little use in World War I and their significance at sea was debated. When the U.S. Navy's Bureau of Aeronautics proposed that the remote-controlled drone could simulate aerial attacks on ships, what exactly this *meant* was not yet known; the bureau hoped that the target drone would not only inform anti-aircraft practices but also settle the question of whether battleships were vulnerable to aerial attack.

In the United States, William "Billy" Mitchell is credited with promoting the concept of air power. He served as deputy commander of Air Aviation during World War I, after which he advocated for an independent air force as assistant chief of the Army Air Service. During the summer of 1921, at Mitchell's urging, a joint army–navy exercise known as "Plan B" was organized to simulate aerial attacks against ships. The results of the exercise were disputed and continue to be today. Among the best-known trials was a test bombing of a captured German battleship from World War I, the *Ostfriesland*, which was also a news media event. The goal was to mimic air-to-sea warfare by attacking the battleship with aerial bombers. Though the tests claimed to replicate conditions of war, this was an overstatement. The *Ostfriesland* was a static target and, unlike a battleship in war, could not perform any maneuvers. On July 20, army, navy, and marine corps aircraft ran the first wave of attacks, which were halted due to poor weather. Only smaller munitions were used, and unsurprisingly, the damage was not substantial. The following day, in another wave of attacks—this time with larger bombs—the *Ostfriesland* eventually sunk. Following this mixed show of aerial might, the demonstration bombings against the *Ostfriesland* failed to create a consensus within the military about air power.

Two articles in the *New York Times* suggest diverging interpretations of the *Ostfriesland's* sinking in the media. After the first day, the *Times* ran the headline, "Bombs Fail to Sink the *Ostfriesland*," reporting that "Fifty-two bombs, weighing 18,990 pounds, loaded with heavy charges of TNT were dropped, and of these thirteen bombs, weighing 4,470 pounds, fell on the deck of the *Ostfriesland*. Only four of the thirteen bombs making these direct hits exploded."[8] Far from suggesting the primacy of air power, the first part of the trial suggested its limitations. The following day, a new headline announced, "Sinking the *Ostfriesland*." After the eventual destruction of the battleship, the article claimed, "Brig. Gen. William Mitchell's dictum that 'the air force will constitute the first line of defense of this country' no longer seems fanciful." The article went on to support Mitchell's assertion, stating, "A nation unequipped to concentrate her whole air force over the water, if the decision lies there, can just as well leave her navies tied up to the wharves instead of sending them out to certain destruction against a hostile country equipped for this purpose."[9] In this case, air power is tested not only by the military but also as news of the experiments circulated publicly to organize the theater of war proposed by aerial bombardment. Reports from the bombing of the *Ostfriesland* position the trial as a test for the nation, extending the significance of the scene to the American public.

The trial bombing against the *Ostfriesland* was one of multiple tests that aimed to assess the air-to-sea war and occurred in the midst of an escalating antagonism between Mitchell and the navy. Earlier that year, Mitchell had testified to Congress about the vulnerability of ships to aerial attack, questioning plans to invest in a conventional navy. He advocated for the development of an Air Service, separate from the army and navy. In the aftermath, Mitchell was asked to resign by the chief of the Army Air Service, Charles Menoher. Mitchell did not retire, however. A new chief of the Army Air Service, Mason Patrick, sent Mitchell on an inspection tour of Europe in 1921–1922. Mitchell was eventually demoted and transferred to San Antonio, Texas, in 1925. That year, Mitchell was court-martialed for publicly accusing senior army and navy leaders of incompetence following the crash of the helium ship *Shenandoah* in September 1925, which killed fourteen crewmembers. Mitchell was charged with discrediting the military service and found guilty. He was suspended from active duty for five years and resigned from the military in 1926. While he never returned to the service, he continued to be a vocal advocate for air power until he died in 1936. His downfall and resignation indicate the precarity of air power as a concept in the interwar period and the extent to which air power's success was not a foregone conclusion, but rather hotly contested even among the military elite. The tactics of aerial war, which Project Drone in the next decade was supposed to assess, were uncertain.

It is common to read the rise of air power in the twentieth century as part of a larger cultural trend toward dehumanization. Such an approach emphasizes the affective indifference of distant bombing, its speed, and its mechanization. Critics often link these transformations to the rise of technoscience within the military. Carl Schmitt, writing in the aftermath of World War II, posits air power as a new spatial order organized by technological domination. He writes, "Above and below . . . can be thought of only naively, from the perspective of an observer who, from the surface of land or sea, looks up and down, up and down, while bombers pass in the airspace overhead and execute their missions from the sky to the earth."[10] We can think of it "only naively," as the disorientation created by aerial bombardment undoes the relationality between the two sides on the battlefield and instead creates a form of technological domination unaccountable to what is below. A vertical form of domination premised on targeting from above replaces a horizontal battlefield, extending the indiscriminate ability to attack to the land and flattening earlier distinctions between war on land and sea. Schmitt's theory takes up the contested claim made by Mitchell in the interwar period as the basis of a new global order.

A further consequence of the disorientation of aerial war is that any attack is rendered justifiable. Schmitt explains that under vertical control the "victors consider their superiority in weaponry to be an indication of their *justa causa* . . . [and] the discriminatory concept of the enemy as criminal and the attendant implication of *justa causa* run parallel to the intensification of the means of destruction and the disorientation of theaters of war."[11] Schmitt thought the global order organized by air power, and thus untethered from the land below, would lead to the intensification of war. With the rise of air power, all limits that had been previously placed on war would be undone, resulting in a world order where might is right. The concept of the "just enemy" in particular would unravel, replaced by the enemy-as-criminal rather than as equal opponent.

Air power, for Schmitt, would be fueled by escalating technological superiority, which itself would be metonymically reinscribed as a more general political and civilizational superiority: we are *above* and you are *below*, says this logic, therefore we are superior—"above you," as one would say in English. The enemy had already been redefined to include the colonial other as criminal prior to World War II; however, modern European conflicts retained the concept of the just enemy, now undone by aerial war in Schmitt's view. Schmitt's claims fail to account for the possibility that the tectonic shifts reorganizing the global order after World War II cannot be fully explained by advances in weaponry alone. The nascent technologies of air power, after all, failed more often than they succeeded. Right, in other words, is not tied to actual might.

Rather, the above/below ideology of civilizational superiority continued to function even in the absence of effective aerial weaponry. Air power thus suggested a framework that named the enemy as criminal even before aircraft were "superior." Priya Satia outlines how in the early twentieth century British aerial bombing campaigns and drone experiments aimed to control colonies in the Middle East; air war was not an extension of ground war in these contexts, but a form of occupation.[12] As with early experiments to test air power, there is little indication these campaigns achieved their intended effects. Rather, they suggest how the emerging theater of aerial war already positioned the enemy target as criminal other, even while the significance of aerial bombardment within military institutions was hotly contested and it was not yet a "superior" technological practice.

A less commonly cited example of air power from 1921 extends the enemy-as-criminal framing to extraordinary extralegal procedures internal to the United States. A month after the demonstration bombing against the *Ost-friesland*, the U.S. Army was called to intervene in a labor dispute between coal miners and coal mine operators at the behest of the West Virginia state government. The federalization of the National Guard for World War I allowed Secretary of War Newton Baker to implement extraordinary measures that bypassed the law and allowed for direct, internal military interventions. This loophole remained in place even after troops returned from overseas and was used in Mingo County between 1919 and 1921. More than five thousand miners mobilized in support of their union, while a local militia of a similar size organized against them. Clayton Laurie explains, "In labor disputes . . . federal regulars were sent in to suppress what were deemed by local, state and federal officials as radical and foreign-inspired labor uprisings and challenges to legally constituted civil authority."[13] Ultimately, the use of aircraft in Mingo County would tie air power to the control of civilian populations in labor disputes, thus linking air war with domestic control.

On August 26, 1921, under the orders of the deputy chief of staff and the chief of the Army Air Service, Mitchell flew to Kanawha Field, outside Charleston, to assess its suitability for reconnaissance or tactical air support while the U.S. government weighed its options. "Upon landing, Mitchell, never one to mince words about airpower, commented to the press that the Army Air Service, by itself, could end the civil disturbance by dropping canisters of tear gas upon the miners. If that failed he recommended the use of artillery by the ground forces to bring the crisis to a speedy conclusion."[14] Mitchell spent only a day at Kanawha Field and did not have the chance to deploy the tactics he promoted. However, when President Warren G. Harding did decide to intervene, only copies of his proclamation—not tear gas canisters—were dropped by private aircraft on September 1, 1921. He also decided to close the legal exception that had enabled federal troops to be deployed domestically at a state's request.

Meanwhile, coal companies hired aircraft to fly over the miners and drop homemade bombs filled with nails and metal fragments; the bombs missed their targets or failed to explode. The same day, twenty-one army aircraft were sent to West Virginia, though only fourteen arrived as a result of mechanical failure and inclement weather. Mitchell was ordered to stay in Washington, D.C. The aircraft never used their armament, though "they performed several reconnaissance missions and enjoyed the unique distinction of being the first air unit to participate in an American civil disturbance."[15]

Federal troops occupied Mingo County for three months until the stand-off between the miners and the local militia eventually ended. Although he was not there, Mitchell subsequently claimed the "Mingo War" was a paragon of air power and demonstrated its ability to control across land and sea. Coal mine owners, borrowing from the imaginary, attempted to drop homemade bombs on the miners. Yet in the end, it was not air power *as a new form of technology* that thwarted the union's strike but, rather, the combined institutional force of local, state, and national government, enabled by the extralegal procedure that permitted U.S. military forces to be used on domestic soil. In other words, air power only has power to the extent that it is given that power by an institutional network of powers.

Schmitt argues that aerial attack introduces a third dimension of war through the vastness of air space and its potential to exert global control, if mastered. Mitchell similarly claimed the dominance of air over the land and sea. In the interwar period, these theories circulated globally and shaped military developments that led to the massive bombing campaigns of World War II and the enormous loss of life that came as a consequence. Tami Biddle emphasizes that the domination imagined by interwar air power proponents was overstated, even as their ideas eventually took hold in the U.S. military. She highlights two failures of air power theories: First, aircraft were vulnerable to anti-aircraft defenses and bombing was much less accurate than expected. Second, civilians on the ground were largely able to withstand aerial bombardment despite its devastation. The interwar assumptions of air power are, according to Biddle, based in rhetoric, not fact.[16]

By the 1930s, Mitchell's status as a disgraced antagonist to the U.S. Navy had shifted. As Williamson Murray notes, "Navy reformers . . . found Mitchell as [a] useful foil for pushing the navy's leadership towards serious investment in naval air power."[17] As the world raced toward World War II, Project Drone at once tested and defined the as-yet murky future that investment in air power would bring.

Human-Media-Machine: The Drone as Network

Remote-controlled flight in the 1930s was not developed as a replacement for piloted aircraft. Instead, it proposed to test the U.S. Navy's anti-aircraft defenses and mimic aerial attack. Project Drone aimed to maintain the remote operator's control over the plane, even as actions and emotions associated with the project overlay, confuse, and disavow interconnections between the control pilot and drone. The navy's remote-controlled target plane was linked not only to the changing context of air war I discussed in the previous section but also to a broad range of innovations with radio and telephone, the uses of which extended beyond the military into homes throughout the United States.[18] Radio signals meshed with air space, offering a way to act at a distance through the discontinuous human and machine parts. Yet, radio did not just communicate between the operator and drone; it also set up their contradiction.

Consider again the scene at Navy Research Laboratory (NRL) mentioned at the outset of the chapter, which dubbed that target plane "drone" based on its insectlike qualities. The recollection suggests the arbitrariness of the codename, though the term persists and resonates. In *Insect Media,* Jussi Parikka proposes an intricate theory, examining how insects are "carriers of intensities and modes of aesthetic, political, economic, and technological thought."[19] By naming the drone after a male bee, engineers at the NRL interlinked radio control with a long history of insect media and the forms of social and political organization it proposes. A hivelike organization emphasizes hierarchy and order, as well as a social association that causes the individual bee to become part of a swarm. The name contains the possibility that the actions of the drone exceed the human, while at the same time defanging any potential threat: a drone, after all, has no stinger. The drone-as-target is thus represented as a kind of technology that is both uncanny and nonthreatening, inhuman yet still safely gendered. In human-machine systems, Lucy Suchman observes, "the distinction between person and machine rests on the traffic back and forth between the two terms that questions of human-machine identity and difference matter." In Project Drone, human, machine, and media parts shape what will come to be a mechanized context for air war, resulting in "the ongoing, *contingent coproduction* of a shared sociomaterial world."[20] However, these exchanges did not always perform as expected.

The drone represents the ideal of the social insect's hive mind, linked and effortlessly functioning in perfect unison; yet, this name was more aspirational than reflective of reality. In the development of the project, the team's engineers reckoned with what would happen if the aircraft lost its connection to the remote operator, who guided the flight. Human and machine control was also figured by its failure. The NRL's "drone" drew on an earlier program to create an aerial torpedo, tested by the navy in the 1920s. When the proposal to build a remote-controlled target reemerged in 1936, some senior navy officers questioned its utility based on a remotely guided seaplane tested as an aerial torpedo. They recalled in their response to the new project a test flight in September 1924, in which a seaplane "was taken off without pilot, flown for about twelve minutes, and landed. The plane sank after landing."[21] A puncture in the seaplane's pontoon was the cause. The radio control gear was retrieved from the submerged aircraft and tested again in October with similarly flawed results. When the researchers attempted another test in December 1924, the plane crashed on takeoff and the project was "allowed to die a natural death."[22]

This account of the failed project reflects Bruno Latour's conception of technology, where we never see "people on the one hand and things on the other." For Latour, technologies are "programs of action, sections of which are endowed to *parts* of humans, while other sections are entrusted to parts of nonhumans."[23] The unmanned but remote-controlled target plane underscores its status as an entanglement of human and machine parts. That entanglement, however, often failed to create the action to which it aspired.

The officer-in-charge of Project Drone, Delmar Fahrney, was an aeronautical engineer trained at the Massachusetts Institute of Technology (MIT) who worked with other researchers at the NRL. The "out of control" aircraft preoccupied the engineers of the new project who sought to show radio control could be used to guide a distant aircraft as if it were piloted by a human. Among the key early considerations for the team building the drone was the possible failure of radio communication. A monthly report outlining a possible design suggested, "If, for example, the receiver gets no signal after an elapsed period of two minutes, the plane is placed in a turn by the time relay. If no signal is received after twenty minutes, the controls are neutralized and set for the landing condition and the throttle cut."[24] The failure of the parts of the drone to respond are just the moments when Latour's program of action comes apart. Latour names these "anti-programs."[25] The drone, however, does not merely acknowledge an anti-program as a mode of organizing human and nonhuman action, but takes that disavowal as constitutive and equal in importance to synthesis. Human, machine, and media parts contradict the organization the drone aims to create.

By March of 1937, the basic parts of the radio-control system—a modulator, demodulator, and a hydro-mechanical system—had been fabricated and tested. The drone drew on telephony, which allowed multiple signals to be sent from a control box, using dial commands, to move different parts of the hydro-mechanical system to operate the aircraft. The first tests involved a manned control aircraft, which sent radio control signals to the "laboratory drone" inside the navy's facilities, monitored by the engineers leading the project (figure 1). The transmitted signal moved the hydraulic servo-valves that would activate the aircraft's actuators, used to maneuver the plane also stabilized by an onboard gyroscope. Early tests showed the radio control to be functional from an aircraft twenty-five miles from the laboratory drone. While the NRL team developed the radio system, the airframe was built at the Naval Aircraft Factory in Philadelphia.

Anxious to work on a model that could be flown, the team of engineers at the NRL procured a training plane, NT DRONE, for their experiments after construction with the new airframe. NT DRONE (figure 2) had previously been used to train navy pilots and could still be flown by a pilot onboard. Initial tests operating the NT DRONE with radio control used a "safety pilot," a pilot onboard to take over controls of the plane in case the radio controls malfunctioned in flight. Here, human and machine overlay one another, with the pilot onboard if the system malfunctioned. The NT DRONE was initially flown by the safety pilot, Fred Wallace, and the flight shifted to radio control once it was airborne with the pilot still onboard. In his monthly report, Fahrney describes the test: "At 3000 feet the circuits were tested and found O.K. and then the DRONE pilot was ordered to throw in the gear—shortly thereafter, there ensued the most astonishing evolutions which could only be ascribed to a drunken pilot: the DRONE went into wild gyrations to the right

FIG. 1 Laboratory drone, U.S. Naval Research Laboratory, Washington, DC, 1936
Credit: Collected Files of Delmar Fahrney, NARA II

FIG. 2 NT-1 Training plane, Naval Aircraft Factory, Navy Yard, Philadelphia, 1937
Credit: Lt. Cmdr. Fred Wallace Scrapbook, San Diego National Air and Space Museum

and to the left with plenty of climbs and dives mixed in to give Wallace a most harrying ride."[26]

When the remote pilot begins operating the NT DRONE, the result is "astonishing evolutions." Although the safety pilot can still control the aircraft, "its" flight takes Wallace, now passive in its operation, on "wild gyrations to the right and to the left with plenty of climbs and dives mixed in." Action in the report is attributed to the drone, likened to a "drunken pilot," while the safety pilot rides out its antics. Fahrney's report continues, however: "After a few moments of anxious concern it developed that the controls governing climb and dive were satisfactory, but that the aileron controls were decidedly 'hay wire.' . . . The obvious fact that the controls were crossed was not at first apparent because the safety pilot threw out the gear and brought the plane back to level flight after each unusual maneuver."[27] Because wires were crossed, mixing up right and left, a "harrying" first flight resulted. The safety pilot's ability to straighten out the plane actually made it more difficult to see how an error had occurred. The "hay wire" plane suggests how actions moved from the control plane through the radio signal to the drone. When these connections functioned, the drone seems *as if* it could be pilotless. When the signal was crossed, as it was in the first test flight, missed communications led to chaos and confusion over the control of the drone. Of the next attempts, Fahrney reports, "After the aileron control was properly hooked up, following this first awkward flight, the next test hop proved that the radio control was adequate for all normal flight maneuvers."[28] The language used to describe *successful* unmanned flight is mechanical. The faultless drone is "adequate" and "satisfactory"; by contrast it is imperfect and human, operated by a "drunken pilot" when it suddenly reveals its contingent nature. Despite the composite nature of the drone's "drunken" flight, the language splits it into *either* human *or* machine, disavowing the network that makes its flight possible at all.

When the airframe built for the project by the Naval Aircraft Factory finally arrived, Fahrney notes in his report from November 1937, "No further testing is scheduled for the ever faithful little NT 'DRONE' which suffered through many hours of radio controlled mistreatment. While all safety pilots reverently view its passing into the discard, they nevertheless feel relieved that its testing days are over."[29] The passage reverses the suffering of the NT DRONE and the safety pilots; one can only imagine how the safety pilots handled the jostling of "many hours of radio controlled mistreatment." The report emphasizes the obedience of the technology—it is "ever faithful." But it is also safely contained, "little," and thus nonthreatening like its stingless insect namesake. The passage makes a brief stab at humor, commenting on the pilot's "relief" at the drone's "passing into the discard," but undercuts itself with the odd choice of "reverently."[30] To "reverently view its passing," taken together with ascribing its "suffering," suggests a *human* quality to the drone; one watches its passing as one might watch that of a fallen comrade. In this way, Fahrney once again undermines any clear-cut distinction between human pilot and unmanned machine.

Even as Project Drone tried to make this distinction, what is human is not always (or ever) straightforward. By November 1937, successful drone flights led naval engineers to propose a test without a human onboard. In his report, Fahrney recalls the new terminology that results. Describing the preparations for the first flight without a safety pilot, he writes, "When a human pilot climbs into his plane to fly alone without an instructor for the first time, the flight is called a 'Solo.' In this case where a plane flies without a safety pilot for the first time under radio control it is obvious that the term to describe the flight is 'Nolo.' "[31] "Nolo" was first a wordplay on "solo" and, later, an acronym for "no live operator," foregrounding the mechanical operation of the drone. The term continued to be used in the U.S. military throughout the 1950s. Nolo again anthropomorphizes and genders the machine, bestowing the drone with a twist on the honorific used for a pilot climbing into "his" plane alone for the first time. That gendered, humanizing rhetoric at once acts to make the drone seem mechanistic, independent, and beyond human control *while at the same time* undercutting its apparent lack of humanity by granting it human qualities and agency. In fact, the unmanned drone remains under the operation of multiple human operators and entangled with human action at every stage of its mission. "Nolo" frames the machine system through the negation of human action, suggesting how radio and aircraft functioned to produce flight, even as a team of personnel was necessary.

The human operators for the first Nolo flight included Fahrney and the former safety pilot, Wallace, who became a radio controller. Fahrney operated the drone from a ground-control unit for takeoff and landing, while Wallace operated the drone in the air from a piloted aircraft flown by another navy airman. Fahrney explains how they rigged the drone for Nolo flying: "In the safety pilot's seat a special 14 volt aircraft battery was secured which was to supply current for a duplicate receiver and selector which doubled the chances of loyal and faithful operation of the electronic equipment."[32] Here, "loyal and faithful" emphasizes how human attributes are transposed to the machine. This exchange is compounded by the report documenting the test as it describes how "the Officer in Charge opened the DRONE'S throttle by radio and the plane made a normal takeoff. *Very little difference* was noted in the behavior of the DRONE without a Safety Pilot."[33] Once the aircraft was airborne, Wallace took over the radio operation of the drone from a control plane that tailed it. He "controlled the DRONE through simple maneuvers for about ten minutes and then lined up the plane for a well executed landing approach."[34]

Landings and takeoffs were difficult to carry out by radio control. The monthly report suggests the challenges during the first Nolo flight: "As soon as the throttle was cut the DRONE's nose went down abruptly and the front wheel struck the ground before an elevator correction could be applied—the front wheel carried away and the plane slid along for forty feet on its nose as the rear wheels folded slowly."[35] Despite an otherwise normal flight, the crash landing emphasized the struggle of achieving the illusion of perfect, human-independent control. The newly networked system did not yet perform as expected. If a safety pilot had been onboard the drone, he would have corrected for the plane's nosedive. Yet, with the loss of the throttle power, the plane dipped into a crash landing. For subsequent Nolo flights, the remote operator stalled the aircraft before guiding it in for a landing to prevent the drop in the aircraft's nose.

Early test flights with drone aircraft interconnect human operator and remote-controlled aircraft, even as one is defined against the other. Human, machine, and media make drone aircraft, while simultaneously establishing a disavowal of what is human. Gilbert Simondon writes, "The opposition drawn between culture and technology, between man and machine, is false and without ground."[36] Nonetheless, resistances between people and machines characterize sociotechnical relations, not just the smooth programmed action or utility of the system. He contends, "The machine is a stranger to us; it is a stranger in which what is human is locked in, unrecognized, materialized and enslaved, but human nonetheless."[37] Today, the actions of the weaponized drone are made strange and inhuman. Early drone experiments highlight, however, how this characterization was made by interlinking aircraft and operator. In other words, the first Nolo flights were an *extension* of manned flight, rather than an ontological break with it. To make this point is to highlight that the violent consequences of unmanning cannot be attributed to mere mechanization and technological advance. Contemporary critiques of drone warfare often point to the supposed dehumanization of unmanned flight to account for its moral atrocity. Yet, as the interwar reports show, this discounting of what is human is always already an illusion that overlays the interconnection between human and machine. The conceptual division between pilot and machine nonetheless outlives its specific rendering in Project Drone and the exact directions of its consequent disavowal of the human; while the "faithful" NT DRONE was a target to be destroyed, today's Predator drone destroys its targets.

From Loyal Servant to Enemy: A Proto-History of Aerial War

Derek Gregory's "From a View to a Kill" outlines the process by which the apparatus of aerial targeting, organized by the contemporary unmanned aircraft, assembles an object by identifying it within an environment or milieu. He emphasizes how the drone's target is made through calculation and abstraction, rendering life killable. As Weber's analysis of the dyad of target/targeting outlines, however, this relation did not just make an object: it simultaneously created a context for targeting. It is a theater of war, not just a pilotless plane or point of aim, that emerges through these trials and experiments.

Although the first Nolo flight indicated how the radio-control system built for the drone program could be used to guide the aircraft remotely, doubt remained in the navy about its utility. Fahrney described the ambivalent reactions to Nolo tests in his manuscript: "Even though this test produced good results for the most part, yet it came at a time when a hassle was going on in the Bureau about maintenance funds for the project, so route slip comments were not very favorable, as, 'we should ... disband the "Unit" ... as soon as possible.' Cmdr. Stevens was quick to defend the project with, 'If this Unit is disbanded, what is to happen to further development of radio control? I firmly believe that R.C. has enough possibilities ... to warrant keeping it alive.' "[38] There was widespread concern about the usefulness of the project in the Bureau of Aeronautics, as officials questioned the innovations proposed by radio-controlled aircraft. The "life" of the project and how it tied to the transformation of war was debated by military officials.

The navy positioned Project Drone as a simulation of wartime conditions. On May 21, 1938, Rear Adm. H. R. Stark, senior member of the U.S. Fleet Permanent Anti-Aircraft Board, wrote to Delmar Fahrney, officer-in-charge of radio-controlled aircraft, outlining the significance he saw for the project: "The most important use to which DRONES may be put is to determine the effectiveness of our present anti-aircraft armament—the use of DRONES will give us a test that is nearer to wartime conditions than any we have had to date."[39] These comments anticipate the move of the drone unit to San Diego over the summer of 1938 for the next phase of the project: using pilotless aircraft to train anti-aircraft gunners for air war. Fahrney notes in a letter to Cmdr. Albert G. Noble, the fleet gunnery officer, "Fleet Training and the Bureau of Aeronautics are withholding any decision as to future work on this project, pending the outcome of these experimental firings and dependent on the reaction of the Fleet Gunnery Officers as to the value of this type of training."[40] The goal and value of target-drone training were to make sense of the indeterminate context of air war, which also proved a test for the drone itself. While the setup for the navy's targeting trials proposed a mechanical field of war that pit gunner against the drone, these practices were also predicated on uncertainty.

Drone control, as outlined in the previous section, moved between multiple points on the ground and in the air, connected by radio. In attempting to achieve machinelike flight, what emerges is not a replacement for the human pilot, but human, media, and machine parts acting together as an air target. They make a milieu for air war defined as a network, not simply a vertical plane. The target drone transported to San Diego was operated through "beep control," a radio control box that sent nine commands to the aircraft by a telephone dial, a predecessor to a joystick system. The drones were guided by a radio pilot based out of a Chevrolet truck that served as a field station (figure 3). Once airborne, control shifted to another radio operator with a control unit onboard a TG-2 plane following the drone (figure 4). This operator maneuvered the pilotless plane from approximately one mile away. The remote pilot could change the direction and pitch of the plane and operate the throttle and a wing stabilization system. Distance between the aircraft and the drone was limited to the control pilot's sight. At landing, control would again shift to the radio pilot at the field unit.[41] Drone flight interlinked operator, multiple aircraft, and ground control, fitting the drone target into a more complex assemblage that anticipates networked warfare.

Yet, trials were marked not by the ascendancy of a new framework for war but by challenges for both anti-aircraft gunners and the drone target. The first target practice using radio-controlled drones was held on August 24, 1938. Outlining plans before the trial, Fahrney writes, "The first practice will simulate an attack directed by a bomber on the Fleet center and the U.S.S. *Ranger* will take station as a protective vessel. . . . Firing will be opened after the drone passes overhead."[42] The drone ran a rehearsal pass over the ship and then, on the second pass, was targeted by the starboard battery and the port battery.

FIG. 3 Radio control set-up for drone practice at Otay Mesa, San Diego, CA, n.d.
Credit: Lt. Cmdr. Fred Wallace Scrapbook, San Diego National Air and Space Museum

FIG. 4 Control aircraft in flight with drone, n.d.
Credit: Lt. Cmdr. Fred Wallace Scrapbook, San Diego National Air and Space Museum

Fahrney observes, "After fire was opened the target [DRONE] turned right and, as seen from the above ranges, maintained a nearly constant range from the firing ship. From the firing ship, bursts appeared to be to the left of the target. . . . The bursts followed the changes of course of the target but lagged so far behind that it is not believed any hits were obtained."[43] The lag between the drone's maneuvers and the bursts of fire underscored the military's lack of preparation for what would become aerial war. The commander of the U.S.S. *Ranger*, John S. McCain, noted that prior training sessions that used sleeves to trail behind the aircraft, anti-aircraft gunners had a high level of success: "An examination of the sleeves indicated that the starboard battery had three hits . . . and the port battery made four hits . . . RANGER's fire control party was well trained."[44] The sleeve-targets favored the ability to shoot at a target run in a set pattern, following the path of the aircraft that towed them. This reliance on patterned, predictable targets shows up again in McCain's account of the drone's first pass: "On the approach the Drone was noted to be maneuvering and the steady course had not been maintained as planned." McCain insisted that had there been a squadron of bombers, there would not have been such variation, and the battery "would undoubtedly have had planes hit and the formation broken up thus preventing an effective bombing attack."[45] Yet, none of the firings even came close to the drone and, to McCain's surprise, the anti-aircraft gunners on the U.S.S. *Ranger* failed at what was supposed to be a simple aerial maneuver to counter. The challenge of the drone tested anti-aircraft technology, thus shaping the field of aerial battle before its ascendancy in World War II.

As engineers improved drone technology, anti-aircraft training simultaneously shifted to defend against it, but it did so in a patchy, incoherent fashion. McCain's report describes the anti-aircraft gunners onboard the *Ranger*: "From the firing ship, bursts appeared to be to the left of the target and were out of the field of the Range-Finder Operator's glass thus preventing him from supplying information to the Range-Keeper Operators as to the position of the bursts. During this run changes of target angle were given to the Range-Keeper Operators who adjusted the range-keeper set-up accordingly."[46] Teams operating the gunnery would have included a number of personnel, with five to seven men assigned to each gun. The range finder would figure the target speed and altitude, while the range keeper was responsible for a mechanical computer that would predict future target positions. Together, they would establish a firing sequence for the gun. In the first drone trials, this information would have been preset, as the speed of the drone and the altitude at which the aircraft was flying were determined by the commanders beforehand—although unexpected deviations still occurred, as evidenced in this case. The trial of anti-aircraft gunnery aimed to introduce calculation and rationality into the unpredictable scene of attack.

Yet, the next trial run suggests how the networked parts of the drone might come undone, crashing the drone target. McCain's report states, "Anti-aircraft training received from firing on a target similar to a Drone is the *most valuable and instructive* firing that any ship equipped with an anti-aircraft battery can have."[47] This approval takes on further weight given the disastrous outcome of the second firing run. After the first pass, the commanders agreed to open more distance between the control plane and the drone. As in the previous trial, the bursts of fire were distant from the target, the closest coming within 300 feet. Fahrney observes that at the end of the run, "The DRONE flew a divergent course of 10 to 15 degrees from the control plane thereby opening the distance between them still more. Before the control plane could close in on the DRONE at 'cease fire,' the smoke from the bursts obscured the DRONE and the distance was so great that it was impossible to ascertain the flying altitude of the DRONE."[48] In the ensuing moments, the target plane lost altitude, and shifting control between the ground operator and the radio pilot in the control plane, the team lost contact with the aircraft. McCain reports, "Shortly after the last shot was fired the target plane was seen to make a sharp turn to right finally going into a spin and crashing into the water."[49] The drone sank in five hundred fathoms of water. It is unclear in the reports whether the gunners actually hit the drone or if it suffered technical failure. Only a gasoline tank and some small debris were recovered from the wreckage of this supposedly "most valuable and instructive" experience.

Over two years of experimental development and training had gone into building the drone target, only to have the system crash on its second pass over the U.S.S. *Ranger*. Despite this, the simulation is declared "valuable and instructive," and taken as proof of what some naval personnel suspected—ships were more vulnerable to aerial attack than had been previously thought. In his manuscript, Fahrney quotes from a review of the project by his superior in the Bureau of Aeronautics, Claude Bloch. Bloch emphasizes how the conditions presented by the drone presented a problem "never before experienced" based on the target's mobility and concludes, "It is feared that the fixed conditions of speed, course, and altitude of antiaircraft sleeve targets in formal gunnery firings have resulted in control methods *which may not prove sufficiently elastic* for firing effectively on hostile aircraft free to maneuver."[50] Naval personnel counted on a fixed path for the aerial flight, even as the trials pointed to the possibility that aerial warfare would instead be marked by mobility, elasticity, and surprise. The drone's contradictory attributes—especially its variability and lack of control in the air—shaped the U.S. Navy's enactment of the as-yet hypothetical conditions of aerial warfare.

Yet, not everyone in the U.S. Navy was convinced by the tests or saw the drone as a realistic training tool. A few weeks after the tests held by the U.S.S. *Ranger*, a drone was used to mimic a dive-bombing run against the U.S.S. *Utah*. To create a dive-bombing plane, three controllers maneuvered the drone target: the radio pilot on the control plane, an operator at the airfield, and another aboard the ship. The radio control would shift from the airfield to the control plane, while the radio pilot on deck of the U.S.S. *Utah* would guide the target drone on the final part of its forty-five-degree dive toward the ship. After it had been fired upon, the flight angle would be straightened out and the drone target would return to the airfield for landing.[51] Summing up the pass, Cmdr. Walter E. Brown of the U.S.S. *Utah* writes, "On the first run, while firing target ammunition the plane was hit causing it to go out of control and it later crashed 1000 yards port beam. It was not salvaged." His overall assessment of the trials is less than positive: "Due to the cost of the plane and the danger to personnel if the plane goes out of control, it is doubtful this is an altogether feasible program." Brown expresses confidence in the navy's current anti-aircraft measures, describing advocates of the training program as "mistaken."[52] While Brown's views were not enough to cancel the project, they point to an overall confidence within the U.S. Navy in its preexisting anti-aircraft measures.

Recalling the first trials in his manuscript, Fahrney writes, "There were mixed feelings of pessimism and optimism, lament for the failure to hit the DRONE on a high altitude bombing run in the RANGER practice and joy over the crushing defeat handed a diving DRONE in the UTAH practice."[53] Air warfare is typically described as abstract, mechanical, and increasingly defined by technoscientific achievements. Yet, accounts of the interwar drone are rich with affect, from the earlier portrayal of the drone as "loyal" and "faithful" to the drone as enemy. With the drone positioned as the enemy, Fahrney "laments" the drone's success against the U.S.S. *Ranger* and proclaims its "crushing defeat" in the trials with the U.S.S. *Utah*. The structure of the trial transforms Fahrney's previously glowing accounts of the aircraft to instead align his analysis with the pessimism and optimism of the navy gunners onboard the ships. This shift recalls Schmitt's characterization of the indiscriminate quality of aerial war, rendering anyone "below" a target; yet, the drone becomes an enemy not through its superior weaponry but as it emerges through a theater of war.

The target drones continued to be used between 1938 and 1941 to simulate aerial attack. A memorandum from April 3, 1939, describes the training and emphasizes the navy's unpreparedness, again against the U.S.S. *Utah*: "The 1.1-inch battery of the U.S.S. *Utah* was exercised in experimental firing runs on a diving DRONE on Wednesday, 29 March and on Thursday, 30 March 1939. . . . A total of 1500 rounds of service ammunition and 500 rounds of target practice ammunition were expended. Careful examination of the two DRONES used, after the practices, *gave no evidence of hits*."[54] Over 2,000 rounds of ammunition were fired by anti-aircraft gunners against the drone without a hit. In July of 1939, Adm. W. D. Leahy, then chief of naval operations, writes, "Firings against DRONE targets during the past year have demonstrated the ineffectiveness of present control methods and procedures, and possibly, equipment, to combat realistic bombing attacks. . . . The early solution of the AA [anti-aircraft] Defense problem is considered urgent."[55] A special committee reviewed the records of anti-aircraft firings against drones in 1940 and made a series of recommendations. They write, "It is considered the unanimous opinion of the Board that aircraft progress has rendered our AA batteries ineffective against determined attack by modern aircraft."[56] The report goes on to observe that confidence in the navy's defenses against aerial strikes was a "dangerous misconception."[57] Despite these recommendations, the simulated flights—and the vulnerability they exposed—resulted in no immediate changes to anti-aircraft defense systems. After Pearl Harbor, however, the debate over the effectiveness of air war against ships would take a new, deadly direction.

Pearl Harbor

On December 7, 1941, the attack on Pearl Harbor by the Imperial Japanese Navy confirmed the two hypotheses tested by the drone: aerial bombardment was powerful, and the U.S. Navy was inadequately prepared for it. The

Japanese launched 353 fighters, bombers, and torpedo planes from two aircraft carriers. Japanese forces sunk or severely damaged 19 ships, destroyed 188 aircraft, and killed 2,402 personnel. During the attacks, navy anti-aircraft gunners shot down only 29 of the 353 aircraft. The realities of aerial warfare were now disastrously present and material for the U.S. Navy. The wreckage of the U.S.S. *Utah*, the same ship used for experimental firing exercises with target drones in 1939, remains at Pearl Harbor today, a memorial to personnel onboard. While Pearl Harbor is typically presented as a surprise attack that the United States could neither have imagined nor prevented, the interwar drone trials uncover a new history of imagined, aerial warfare.

Air power advocate William Mitchell *had* imagined an attack against Pearl Harbor from the Pacific (although he proposed it would be launched from land) in his 1925 book *Winged Defense*. Intelligence that the Japanese were planning an attack had been sent to President Franklin D. Roosevelt and early radar picked up signals of the aircraft. Hindsight makes it possible to construct a story that seemingly reveals the "truth" hidden in plain sight: military vessels were vulnerable to aerial attack. Yet the U.S. Navy's lack of action prior to the attack reflected the confidence held by senior officials that earlier patterns of war would continue unchanged. Pearl Harbor emphasizes that "actual" conditions of war are unpredictable, defying attempts to be made rational and knowable, despite the three years of experimental trials that had tested U.S. Navy defenses against target drones.

By the end of World War II, improvements in anti-aircraft defenses meant ships became much better at defending against aerial attacks, and the U.S. Navy continued to consider its naval strength to be superior to aerial attack. This narrative is tidy enough, as historical narratives go. But it exists only as a superimposed, retrospective claim. In the interwar period, a small number of U.S. military minds imagined the possibility of air power as total domination over land, sea, and air. In service of that vision, they piled onto the drone their hopes for an airborne avatar that would at once possess insectlike obedience, anthropomorphized loyalty and faithfulness, and mechanized power. But despite these hopes, the drone crashed, sank, exploded, sputtered, and, more often than not, failed at its intended purpose. Simultaneously, it laid the conceptual groundwork for the rapid expansion of anti-aircraft defenses that would fall into place after Pearl Harbor. In other words, there is no smooth narrative of progress here: the evolution of aerial warfare was never a foregone conclusion but a constructed one. There are only disjointed historical histories of targets and targeting that we connect, partially, in hindsight.

2

American Kamikaze

● ●

Forty years after Project Option, a thirty-day, top-secret test mission that deployed television-guided drone weapons in the Solomon Islands in 1944, James J. Hall self-published a book about his participation in the project. The book was called *American Kamikaze*. In it, Hall attempts to recuperate the significance of Special Task Air Group One (STAG-1). Its mission was to bring the target drone of chapter 1 into the television age. In these trials, the drone was designed as a prototype guided missile, transforming the remote-controlled target plane to an assault weapon. Vannevar Bush, head of the United States Office of Scientific Research and Development during World War II, labeled the experiment to develop a television-guided weapon a "debacle."[1] Ballistic missiles took precedence for the American military in the postwar period and efforts to deploy radio and television communication for targeted attacks were ended by the military. By 1984, however, when Hall published *American Kamikaze*, the U.S. military was revising the status of the failed project, ultimately declassifying it and declaring the system a precursor to networked war. In 1990, a letter from the secretary of the navy proclaimed, "On 27 September 1944, a TDR-1 Assault Drone launched and staged a combat attack against an enemy target; the success of this first true guided missile marked a new era in modern warfare."[2]

Hall's title, *American Kamikaze*, explicitly positions the American drone as a suicide bomber. The book retells his experiences in World War II in the third person, relaying the exploits of his unit, relocated from Oklahoma to Michigan to California between 1942 and 1944, before finally being deployed to the South Pacific, and their disbanding after the month-long experiment. He reminisces about being told about the top-secret mission for which they had been selected, recalling the lieutenant saying, "We are going to mount radio-controlled pilotless drones against the enemy."[3] The meaning of "American Kamikaze" receives almost no attention in Hall's account, however. It acts as a haunted, empty signifier, standing for a disavowal that is a counterpart to the technological innovation the navy celebrates in 1990. This chapter takes up that haunted alignment between the kamikaze and the drone, examining how American racial stereotypes tie to a claim of technological superiority through a project dismissed as a failure.

The title *American Kamikaze* reveals how the U.S. Navy's Project Option compares drone and kamikaze, using television transmission to produce an overlay between suicide bombing and drone missiles. The previous chapter showed how the U.S. Navy's drone target shaped the conditions of air war it was supposed to mimic. In this chapter, the Pacific Ocean theater of World War II acts as a framework for the drone, now a weapon rather than a target. While the U.S. Navy purportedly copied Japanese kamikaze tactics with the refashioned target drone, these claims are based on U.S. cultural presuppositions and assumptions. The television-guided weapon is made at once foreign and machinelike in the military rhetoric of the time, but years later, memoirs like *American Kamikaze* reveal the human intimacies and attachments behind the project. This tension between human and machine is enmeshed with the military's use of television; the "actual view" onscreen is also multiple, contradictory, and flawed.

In this chapter, I consider how lethal action is made technological by the television-guided assault drone and how its target is rendered an object. Operator and enemy are disavowed as human or even as *alive* through this technological framing. In *Frames of War*, Judith Butler examines the conditions that make subjects. She writes, "If certain lives do not qualify as lives or are, from the start, not conceivable as lives within certain epistemological frames, then these lives are never lived nor lost in the full sense."[4] Life is conceived through its perception as such within a specific context. Butler explains, "[The] normative conditions for the production of the subject produce an historically contingent ontology, such that our very capacity to discern and name the 'being' of the subject is dependent on norms that facilitate that recognition. At the same time it would be a mistake to understand the operation of norms as deterministic."[5] I take up the opposite, but corollary, question in this analysis: How are conditions produced to disavow the human subject, rendering instead a technology as alive or dead? As with the conditions that Butler describes as facilitating a subject's recognition, the conditions that produce the drone as mere technology are historically situated and by no means given.

Unmanning alerts us to how the production of a subject is also entangled with the discursive and material opposition between human and technology. The epistemic frame of war organized by the human, machine, and media parts of the drone makes everything a potential point of aim, named through the interplay of target/targeting and the totality imagined by aerial domination. Yet, the framework never completely achieves its effect. Butler observes, "Production [of the subject] is partial and is, indeed, perpetually haunted by its ontologically uncertain double. Indeed, every normative instance is shadowed by *its own failure*, and very often that failure assumes a figural form. The figure lays claim to no certain ontological status, and though it can be apprehended as 'living,' it is not always recognized as life."[6] Reworking this shadowlike failure points to the negative of unmanning—that is, what is human, as its double. The very existence of drone aircraft proposes a seamlessly mechanical air war; yet, television static or mechanical failure repeatedly ruptures this picture of unmanned superiority. Butler contends, "What happens when a frame breaks within itself is that a taken-for-granted reality is called into question, exposing the orchestrating designs of the authority who sought to control the frame."[7] Thus, the material failures (static, guidance system failure) expose the cracks in the epistemological frame posited by the drone.

The ontology of unmanning is marked by the repeated negation of what is human, even as, over and over, the human, media, and machine parts exceed the reality they are supposed to produce. The drone is never not human. The drone and the scenes of war it shapes and is shaped by come apart in contradictions, accidents, and contingencies. Two entangled narratives of World War II meet at the television-controlled drone and, in the process, illustrate its overdetermined ontology: the intersection between the racialized status of the "kamikaze bomber" in the American imagination and a history of technological progress. The two narratives rest uneasily with each other. The first shows how racial stereotypes widely used by the United States in its war against Japan fit with the television-guided drone. Virulent propaganda, Japanese internment in the United States, and the atomic bombings of Hiroshima and Nagasaki in Japan cohere in the stories the United States told itself about the drone during the war. These narratives layer over the racial stereotypes implicit in the drone's development with the claim of technological innovation.

By reading both retrospective and contemporary drone mythologies against the historical record, neither drone "superiority" nor the project's "success" makes sense. The television-guided weapon was deployed for a small number of bombings and its significance to the war effort at the time was considered unremarkable. If the project did receive attention, it was because the camera and monitor used in the weapon became an enormous commercial success with the rise of broadcast television after the war. Only in retrospect

could the drone be seen as the marker of a historical watershed. What resonates are the shadows of the drone's failure and the negations that crack the
façade of an inhuman targeting machine, thus troubling the mythos claimed
for that machine.

Television and Our "Electric Eye" of War

In announcing the STAG-1's television control unit, Hall recalls that the
lieutenant's explanation was interrupted by the immediate response of the
personnel. " 'Television?' the buzzing went through the crowd. 'Men! Men!
Yes, television, both transmitting and receiving.' "[8] The reaction indicates the
novelty and indeterminacy of what television was in the early 1940s. William
Uricchio characterizes early television through "interpretative flexibility," a
term used by science and technology studies to describe the malleability of a
technical system, particularly in the early stages of its innovation. Television
"was variously understood as domestic like radio, public like film, or person-
to-person like the telephone." Although the concept of television had been
explored worldwide since the 1920s, the use of television in assault drones
during World War II predated the widespread development of commercial
broadcast television in the United States. Television was highly anticipated,
but exactly what it was or how it would operate was as yet unknown. Uricchio explains that prior to the 1950s "television . . . drew upon journalistic,
theatrical, and (documentary) filmmaking practices," and he argues that contemporary transformations in television "are not so much new as reminders
of the medium's long-term flexibility."[9] Television-guided drones fit comfort-
ably within the multiplication of possibilities conceived for television's image
transmission system.

Longstanding ties between RCA and the military predated the television-
guided weapon. The company was formed as part General Electric in World
War I to create an "all-American" telegraph company as a national security
measure. High-ranking officials in the navy were concerned that telegraph
messages sent through "foreign" telegraph lines, then controlled by Marconi
company, would be intercepted. In the interwar period, RCA became a leader
in commercial radio technologies and broadcasting. An antitrust settlement in
1932 separated the company from General Electric and placed David Sarnoff
in charge. Throughout this period, the U.S. military continued to be a major
client. The army awarded RCA's Sarnoff a brigadier general's star during World
War II for his service to the Signal Corps and he was widely known as "General Sarnoff" thereafter.

Even before taking control of the company, Sarnoff advocated for television. He hired Vladimir Zworykin and billed him as the "Father of Television"
(despite competing patent claims by Philo Farnsworth). Zworykin ran RCA's

television laboratory in Princeton, New Jersey. What television would do was uncertain, however, and among the future uses Zworykin proposed was as an "electric eye" for a remote-controlled weapons system. A memorandum, "Flying Torpedo with an Electric Eye," circulated to RCA on April 25, 1934. The first part of the description describes targeting as a field that could be monitored by television. Zworykin writes, "Television information furnished would be of two kinds, and would be given simultaneously: (1) an actual view of the target which could be sighted by means of crosshairs; (2) accurate information on the readings of instruments in the piloted weapon."[10] The language, like the justification for the target drone project, promotes the objective qualities of the project. Zworykin outlines how personnel could "see through" television to the field of war. He suggests that the innovation of the "electric eye" results in a transparent view of the target. Television would relay "accurate information" and an "actual view," sighted through crosshairs. The act of targeting proposes to overcome distance through a direct view of the battlefield, extending human sight and action.

In the second part of the memorandum, Zworykin focuses on how the pilotless plane exceeds human sight and body. He observes, "Considerable work has been done also on the development of radio-controlled and automatic program-controlled airplanes having in mind their use as flying torpedoes."[11] Radio control alone, however, relies on the operator's vision to direct the missile to its target, limiting the range of the weapon to how far the operator could see. Having established the embodiment of the operator as a problem to be overcome, Zworykin continues, "The solution of the problem evidently was found by the Japanese who, according to newspaper reports, organized a Suicide Corps to control surface and aerial torpedoes."[12] This early claim is significant given that systematic attacks by Japanese kamikazes did not occur until ten years later, in 1944. Zworykin names suicide bombing, at the time an improvised tactic employed by pilots from diverse nationalities to direct a crashing aircraft into a target, as a uniquely Japanese analog to the television-guided torpedo.[13] The characterization proposes a field of war extending beyond the human body and the limits of life. It is defined by total destruction with intent, rather than accident or opportunism, but made through television and remote control rather than the pilot's suicide. Positioning the television-guided drone as an analog to the imagined specter of the kamikaze bomber, moreover, reframes the Pacific Theater as an absolute war where all-out attack is the only possible action.

Zworykin continues, "We hardly can expect to introduce such methods [suicide bombing] in this country, and therefore have to rely on our technical superiority to meet the problem. One possible means of *obtaining practically the same results as the suicide pilot is to provide a radio-controlled torpedo with an electric eye.*"[14] Zworykin claims "technical superiority" would engineer an aircraft that obtained the same *results* as a Suicide Corps, not the *protections* it would provide for military personnel, despite the fact that such protections comprise the primary justification for contemporary drone aircraft. Deadly destruction becomes a technical solution carried out by an "electric eye," and the limits of manned warfare disappear into an engineering problem. Whereas the first section of the memorandum ties television to human sight, the second section lays out the lethality of the torpedo as a series of technical parts. Going on to describe the operation of the television-torpedo, Zworykin explains: "The carrier airplane receives the picture viewed by the torpedo while remaining at an altitude beyond artillery range."[15] In this passage, the torpedo actively "sees" the picture and then transmits it to the receptive aircraft, erasing the role of the human viewer and making the machine's electric eye the active party. The only hint of personnel lies in the side comment that the carrier plane must remain outside the range of anti-aircraft artillery. "Flying Torpedo with an Electric Eye" promotes the weapon as objective and foreign: the torpedo acts as a more-than-human system, disconnected from its operator by technical relations between its parts and the camera.[16] Aerial targeting here plays out as if television reacted to an "actual view" now beyond the human body. Yet, despite its supposedly unmanned, objective nature, when joined with the racially encoded fear of the "Japanese Suicide Corps," the drone also carries out a technological suicide attack. In this way, we return to the tension of the interwar target drone, at once anthropomorphized (here, capable of racially coded suicide) and mechanical (here, via an "electric eye" that provides perfect accuracy in the crosshairs).

Zworykin's "Flying Torpedo with an Electric Eye" suggests a picture of the world as a set of targets organized by machine parts, which are simultaneously racially coded as subhuman. Rey Chow writes, "We may say that in the age of bombing, the world has also been transformed into—is essentially conceived and grasped as—a target. To conceive of the world as a target is to conceive of it as an object to be destroyed."[17] Her argument focuses on "destruction by visibility" exemplified by the atomic bomb, though her argument resonates with the television-controlled weapon. Chow underscores entanglements between racism, war, and knowledge through the bombing of Hiroshima and Nagasaki, outlining how the world organized as target framed Cold War–era area studies. In the postwar period, the United States locates itself in "the position of the bomber, and other cultures always viewed as military and information target fields."[18] Zworykin's proposal foreshadows these arguments, analogizing the American weapon as the technically superior counterpart to the suicide bomber.

The 1934 memorandum describes the "electric eye" as an "accurate view" of the battlefield; this characterization is simultaneously inscribed by a division between "this country" and tactics attributed to "others." The television-guided weapon registers a field of domination marked by technoscience, on the one hand, and Zworykin's typecasting of Japanese tactics, on the other. The memorandum articulates the two registers of violence that Chow emphasizes, defining the target as object and other at once. Internal conflicts about the weapon within the navy only underscore these two registers of violence. Contemporaneous discussions of the televised drone within the U.S. Navy position the United States as a bomber and other cultures as its target, even while insisting on Zworykin's claims that the drone represents "technical superiority"—despite all evidence to the contrary.

Project Option: Drone as Televised Propaganda

As early as 1935, RCA met with representatives from the navy about the possibility of using television to control aerial weapons. The navy invited Zworykin to report on his ideas in 1937. Following his presentation, the review board concluded that they "appreciated the thorough study of television and radio controlled aerial torpedoes, and were satisfied that, at least for the present, the situation does not justify any expenditures of funds for experimental purposes in this field of endeavor."[19] Attitudes toward remotely controlled aircraft shifted as the trials with drone targets expanded, however. In 1939 Commander in Chief of the Navy Claude Bloch wrote the following commentary supporting their possible development: "The extension of the role of the radio controlled airplane from the passive one of a target to the active one as an offensive weapon should be recognized as a reasonable development, and experimentation to determine the most useful field for this weapon is considered fully justified."[20] Bloch's description outlines the drone's changing role, legitimated by the experimental trials with target drones. Yet, within the navy, many were still uninterested in the drone project and opposed its expansion.

The development and later cancellation of the STAG-1 project outline how different groups in the U.S. Navy responded to and shaped the drone project. Donald MacKenzie posits the concept of "interpretative flexibility" to examine nuclear guidance technologies built for missiles from the 1950s onward. Interpretive flexibility provides a useful framework for the drone, where supposedly "technical" innovations are established through multiple practices that entangle social and political interests. In his analysis, MacKenzie shows that missile guidance does "not simply [mean] different things to the different 'inventors,' but also [was] seen by different groups as a solution to quite different problems."[21] His argument does more than enter into debates about whether nuclear missile guidance was accurate—itself a flashpoint of the early 1980s. Instead, he shows how the concept of accuracy itself emerged through specific networks of practice and knowledge that shape and were shaped by guidance technologies; his argument is that precision is created in tandem with the sociotechnical relations that underwrite guidance. He writes, "Take away the institutional structures that support technological change of a particular sort, and it ceases to seem 'natural'—indeed it ceases altogether."[22] Accuracy, as such, is not a teleology measured by ever greater progress in hitting a point of aim but a social and political claim, mobilized by a range of actors in multiple and, at times, incompatible contexts. Similarly, the drone project is marked by varying claims for the system: it is by turns a target plane, a television-guided weapon, and an alternative to manned flight. Its "reasonable development" only occurred when supported by the navy; the moment that institutional support fades, the technical evolution proposed by Bloch comes apart.

From the earliest days of his involvement, Delmar Fahrney saw the possibility that the radio-controlled target begun in 1936 might be a weapon. By 1939, Fahrney was informally organizing support within the navy to build an assault drone and, subsequently, recruited key individuals from the earlier experiments with target planes to participate in the project.[23] In 1939, the same year that RCA demonstrated television transmission at the World's Fair in New York, the company received a contract from the navy to produce an experimental prototype of television control. During World War II, RCA produced thousands of television sets for the project. The refinements to the television tube that were used to guide the drone aircraft led to the development of the image orthicon, which made a clearer onscreen image. This innovation became a crucial part of commercial, mass-produced television sets built through the 1960s, again highlighting interpretive flexibility.[24]

The navy officially began its assault drone program on March 22, 1940, when then-chief of the Bureau of Aeronautics, Ernest King, approved the conversion of a TG-2 aircraft (previously used as a control plane for drone targets) to remote guidance by television and radio control. This was more than a year before the United States entered World War II and King became the chief of Naval Operations. Walter Webster, manager of the Naval Aircraft Factory who oversaw production of the assault drone, described the experimental project a year later. In a report from August 7, 1941, writes: "The DRONE was maintained under continuous radio control, television guided, for a period of forty minutes (during which time the control pilot was not able to see the Drone), made runs on a target, returned the DRONE to the initial point and repeated the runs. The maximum distance that a clear picture was obtained (television) was six miles."[25] Webster emphasizes the distance between the control aircraft and the drone to indicate how the television-guided platform might operate beyond the line of sight of defenders. In September 1941, two additional TG-2 planes were assigned to the project and converted to radio and television control. By November, the Bureau of Aeronautics issued a report that considered production possibilities on a larger scale, looking to obsolete airplanes as possible platforms for the remotely guided weapons, as well as cheaply produced plywood airframes.[26]

With the U.S. declaration of war after the attack on Pearl Harbor on December 7, 1941, the debate over the efficacy of a drone missile intensified. With a large part of its fleet and aircraft destroyed, many within the navy emphasized the importance of rebuilding naval warfare—including further developments with navy aircraft and aircraft carriers—instead of new and unproven systems like the drone. At the same time, others promoted "technological" advantages to counter Japanese forces. Captain Oscar Smith of the Naval Bureau of Ordinance was one such advocate. Apparently unaware of the top-secret developments with radio and television control already underway, Smith wrote on December 15, 1941, "We need no suicide squad to dive torpedo laden airplanes into the sides of the enemy ships. Let a simple type of radio control be placed on a plane, and we have a suicide pilot who will not falter, but will obey all orders of the controlling plane, and will not hesitate to fly within 100 yards [of the enemy ship] before dropping his torpedo."[27] Smith would become the most prominent promoter of the television assault drone program in the U.S. Navy. Much like Zworykin, he sets up a displacement between the human and technological through the tactics of a suicide pilot, personifying the radio-control plane as obedient and unhesitating. It is unclear whether Smith knew about Zworykin's 1934 memorandum; his statements drew on stereotypes that circulated broadly.

The U.S. military stereotyped Japanese forces as engaging in kamikaze tactics even before being systematically attacked by suicide bombers in 1944. This may have been because of Japan's no-surrender policy, as well as accounts of a Japanese pilot who was shot down and crashed his plane into the deck of a ship during the Pearl Harbor attacks.[28] The typecasting of Japanese forces here also underlines how race was used by the United States to portray and organize its enmity with Japan, as well as its own supposed superiority.[29] Smith uses of the notion of the kamikaze to frame the radio control technology he envisions as more-than-human, unfaltering, and compliant in its approach to "death." He equates the drone with enemy suicide tactics, making it strange, even as he proposes remote control as a supposedly more advanced, American way to wage war.

After visiting the project led by Fahrney in February 1942, Smith suggested the development of the assault drone be expedited. In May, the navy laid out a plan to mass-produce a remotely controlled, television-guided weapon. The Bureau of Aeronautics approved a proposal to build between 1,000 to 5,000 systems. John Towers—a navy pilot and proponent of aviation within the navy since World War I, who assumed the role of chief of the Bureau of Aeronautics after the decision was made—was more hesitant. Towers requested that the project develop only five hundred units and be named "Option." He notes, "This bureau is considerably concerned over premature commitments of funds, materials and personnel to this project which otherwise would be available for current needs."[30]

The top-secret Fleet Special Air Task Force, which included the STAG-1 whose exploits Hall recounts in *American Kamikaze*, began training crews to operate the television-guided drone in 1942. Smith was given the new rank of commodore and oversaw the program (although, with his background in the Bureau of Ordinance, he was seen as an interloper within naval aviation). Final proposals called for over 3,000 personnel, 99 control planes, and 891 drones divided into three Special Task Air Groups (STAGs). By early 1943, however, only twelve assault drones had been built by the Naval Aircraft Factory. Although the pilotless planes incorporated television and radio control, they were otherwise low-performance vehicles built of plywood due to the shortage of metal during the war. These assault drones were, in fact, slow and could only be maneuvered in a rudimental way. Further, the cost far exceeded the available budget, adding to concerns about the experimental weapon.

Interstate Aircraft was contracted to build the next model, the TDR-1, which was also made of plywood and tested in late 1943 (figure 5).[31] In practice, the distinction between piloted and pilotless flight was blurry. An onboard pilot ferried the drone to its location, so the TDR-1 had a removable cockpit canopy and the remote controls could be disabled. The television-guided air vehicle was 37 feet and 11 inches in length with a wingspan of 48 feet, 11 inches. The TDR-1 was designed to carry a 2,000-pound bomb and had a maximum speed of 140 mph

FIG. 5 Navy TDR-1 drone during a test flight with human pilot, n.d.
Credit: U.S. Navy

and a range of 426 miles.[32] Wurlitzer Musical Instrument Company, a piano manufacturer, was subcontracted to fabricate the plywood frames. In contrast to the idealized, seamless technical system imagined by the project, the working parts of the television-guided weapon were made under budgetary and material constraints by companies that also produced everyday items.

Once the TDR-1s were built in 1943, the navy still did not deploy them. Towers, now commander of the Pacific Fleet, continued to resist efforts to include the television-guided drone in his battle plans. Gaining in the war, navy commanders were satisfied with the existing tactics and matériel in the South Pacific. The television-guided weapon was untried. Reviewing the project ten years later, Fahrney offers this analysis. He highlights tensions between Smith, who came from the Bureau of Ordinance, and Towers, a pioneering aviator within the U.S. Navy who had been thwarted in his attempts to use aircraft in World War I. "Considerable light can be thrown on the attitude of Towers toward the assault DRONE program if we analyze the personalities involved in this issue concerning its combat employment. Towers was well disposed toward the idea of radio controlled and guided air traversing vehicle for assault usage. . . . He had misgivings, however, based on his experiences with [previous unsuccessful aerial torpedo experiments] and the general conviction that it took a human pilot to fly an air machine. Having been one of the first naval pilots, he was reluctant to concede that an aviator would be displaced by a robot."[33] Smith argued for radio and television control to take up what he calls the enemy's "suicide" tactics. Fahrney conceived of the "robot" system as a logical progression of war. Towers believed an aviator could not be replaced by a pilotless plane. This power struggle informs assault drone technology, military budgets, and the use and ends of technology in wartime, all articulated through a struggle to define "reasonable" technological advancement.

The scale of the assault drone program was reduced significantly in February 1944, and most of the personnel who had been trained for the television-guided assault drone program were reassigned. An ally of Towers, Capt. H. B. Temple, was placed in charge of the navy's guided missile program.[34] That year, Fahrney was reassigned to serve as head of the Logistics Section of the Aircraft Command, the only position he would hold that was not related to drones or guided missile development between 1936 and his retirement in 1950. Commodore Smith continued to exercise some influence in support of the experimental weapon, however. His argument that the television-guided assault drone should be tested in combat held sway with the chief of Naval Operations, Ernest King, who had initially given the go-ahead for the project in 1940. This resulted in the deployment of the one remaining Special Task Air Group in June 1944.[35] By the end of the summer, though, King terminated the experiment before the unit ran any missions. The navy transferred surplus radio and television technologies to the army in an effort to reduce costs. The navy would instead pursue "the latest advances in the science of propulsion, aerodynamics, and electronics,"[36] and future developments would emphasize the strategic advantages of the sea fleet and ballistic missiles.

While senior naval officers debated and ultimately decided to cancel the television-guided weapon, Commodore Smith's second-in-command Robert Jones used the film *Service Test of Assault Drone* to convince navy command-ers in the South Pacific to deploy the weapon. Fahrney describes how "Jones made a flight to the headquarters of Commander Aircraft in the Northern Solomons on Bougainville and conferred with Brig. Gen. Clauss A. Larkin . . . regarding the employment of the guided missiles in strikes against the enemy. After Larkin viewed the films of the [tests] he was convinced suitable targets could be found. Dispatch authority was given . . . for a thirty day trial."[37] Jones had been part of the project since the trials with target drones in 1937. The 1944 film stages an attack with a television-controlled drone against a beached Japanese freighter, the *Yamazuki Maru*. Its wreckage remained from the Battle of Guadalcanal the previous year.

The film *Service Test of Assault Drone* offers a record of the television-guided aircraft, staging both an experimental test and an idealized view of the system. It was screened for military authorities as a final effort to secure support for Project Option, just as the program was being closed by the navy.[38] The film is a documentary of tests carried out by STAG-1 using four drones to attack the wrecked freighter. In the film, orders to the STAG-1 unit are displayed on title cards. Time and targeting organizes the narrative, structuring the film as the "actual view." An intertitle early on indicates the drone strike would occur at "fourteen-hundred hours." The goal of hitting the target on time functions as a marker of the success of the experiment, as well as a cinematic climax for the sequence of images. The staged mission against the *Yamazuki Maru* enacts a "successful" attack against a ship that had already been defeated by conven-tional means, shaping what is imagined as possible. The document can be read as a filmic enactment of Zworykin's "Flying Torpedo with an Electric Eye": it attempts to produce a pilotless, strange counterpart to piloted flight. Yet, the film also emphasizes the immersive connection between the operator and tele-vision through the transmitted image.

The footage from the test begins with a title card indicating that there is no onboard pilot: "The drone in NOLO [no live operator] condition ready for take-off."[39] The TDR-1 drone is then pictured in the center of the frame on an empty runway, palm trees in the distance. None of the radio-operators involved in the TDR-1's takeoff are in the picture. The next intertitle states that each TDR-1 holds a 2,000-pound bomb and is radio-controlled from a TBM plane, as the image pans across the runway showing the other assault drones and island landscape in the background. In the next shot, a sleek aircraft without a cockpit launches from the runway and takes off into the sky; no human appears. Solely

FIG. 6 Film still from *Service Test of an Assault Drone,* nosedive during drone takeoff, 1944
Credit: National Air and Space Museum

in the shot that follows, after the second TDR-1 fails to take off (figure 6), does one glimpse the personnel involved, who rush onscreen to attend to the drone's noseover. It is a result of technical difficulties that the viewer sees the operators.

Once airborne, a title card states, "During attacks, control planes remain *seven miles from the target*." The next image shows the exterior of the control aircraft against the open sky, with no sign of the television controller who is onboard. This shot shows the "carrier plane" that Zworykin described in his proposal; the *control plane itself* ostensibly networks the operations of the drone between the television and the plane, as though the weapon were self-directed. After showing the control aircraft, the next title card sets out the orders: "To crash the side of the breached Jap freighter, *Yamazuki Maru*, Cape Esperance, Guadalcanal, in succession, commencing at 14:00." The following shot is a close-up of the beached freighter deck, panning across the point of aim described in the previous intertitle. More than half of the film is devoted to showing the drones, control plane, and target in succession. The images organize how the drone operates and how it will target, proposing a technological system that leaves out navy personnel integral to the assault drone's functioning (who are only seen during the noseover).

The second part of the film shifts the focus to television transmission by recording images from the receiver in the control plane. Although the

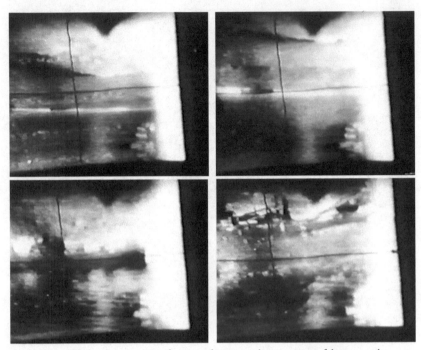

FIG. 7 Film stills from *Service Test of an Assault Drone,* television view of drone attack against a beached Japanese freighter, 1944
Credit: National Air and Space Museum

drone's operator never appears onscreen, the film viewer watches the drone strikes from his perspective, seeing through the camera on the weapon as it dive-bombs toward the beached freighter. In this part of the film, the viewer is invited to see through the television transmission as if he were in the dive-bombing aircraft. The film carefully sets up this perspective; to achieve such a view, the filmmakers would have had to position a movie camera inside the control plane to film the screen of the television transmission. The viewer would see the battlefield at once through immersion with real-time image transmission and from the technological distance set out in the previous section. The title card at the beginning of this sequence states, "At 13:58 control pilot sights target on television screen," using time to indicate how the filmed images on the monitor are the "actual view" of the strike. A grainy television transmission follows the intertitle, showing an almost unintelligible island landscape with the freighter in the foreground (figure 7). The target might not have been recognizable if a prior sequence had not shown a close-up of the deck. Onscreen, for both the pilot watching the television screen and the viewer watching the film, crosshairs indicate the point of attack.

As the drone dives downward, the freighter becomes more prominent in the operator and viewer's screen. The water in front of the *Yamazuki Maru*

FIG. 8 Film stills from *Service Test of an Assault Drone,* landscape view of drone attack against a beached Japanese freighter, 1944
Credit: National Air and Space Museum

glares white with the midday light, and the ship comes to occupy more and more of the frame. Visual noise interrupts the transmission and the display flickers, relaying the simultaneity of the television image. The picture returns and *Yamazuki Maru* fills more of the screen, turning black as the drone crashes into the deck. The intertitle draws the viewer's attention to the connection between the black screen and the completion of the mission: "First drone TDR #860 strikes at scheduled time." The next shot is from the point of view of another camera filming the test against the freighter (figure 8). The assault drone dives into the *Yamazuki Maru,* followed by a large explosion. The two shots establish the impact of the television-guided weapon as it is destroyed when it hits the deck of the ship. The first image disappears with the explosion. The second view, however, shows what has happened through a landscape shot: billowing clouds of smoke from the ship contrast with the tropical island in the background.

The drone explosion obliterates the attacking aircraft and camera, the picture mirroring kamikaze flights that would target U.S. Navy ships in the coming months. In the sequence, the television image aligns viewers with the perspective of the drone and operator, while the second shot disconnects them from this point of view, showing the strike as a technical sequence. Two of the four planes hit the ship, and a final bomb strike closes the film, after which a title card appears: "The End." This film documents and enacts an account of a television-guided missile strike, directing viewers how to watch the drone and creating a particular, contradictory role for the human operator with whom they are aligned. The viewers see through the television lens and watch the bombing from a "neutral" camera recording the strike, while the real-time transmission relayed from the drone's camera zooms in on the target in the crosshairs. The television assault drone targeting the freighter onscreen becomes ever closer before turning black, immersing viewers in the trajectory to the target and its impact. These images point to the operator's effacement—particularly the second shot of the strike, which distances the operator from the view of the television camera, showing the drone diving into the ship from the perspective of an onlooker as if he played no role.

This complex history of the television-guided networked drone complicates the straightforward opposition between the drone and the kamikaze that Grégoire Chamayou makes. Based on a brief analysis of Zworykin's memo, Chamayou contrasts "the suicide bomber who crashes once and for all in a single explosion . . . [and] the drone which fires its missiles repeatedly, as if nothing happened."[40] This characterization leads Chamayou to theorize the "twin" tactics of the drone and the kamikaze, a ghostly machine versus a courageous combatant, both organized in the service of targeting. Chamayou is, of course, correct that there is a stark difference between the death of the kamikaze pilot and the technological assault carried out through the television-guided system. However, Chamayou's formulation overlooks how missions undertaken by the "American kamikaze" functioned in the broader context of war. His contrast between the suicide pilot and the drone misses the set of contradictory contexts established through the development of the navy project and the ambivalence that ultimately marked the use of the assault drone in World War II. The comparison offered by Chamayou equivocates between the first television-guided weapons (each of which was used only once in battle) and contemporary unmanned aircraft (which actually *can* "fire . . . [their] missiles repeatedly, as if nothing happened"). In this way, Chamayou smoothes over the history of human failure, bureaucratic disruption, and mechanical accident that dogs the concept of the drone as an inhuman, lethal weapon.

With the term "ghostly machine," Chamayou displaces the significance of human action that drives the project and the drones themselves. In this way, Chamayou reiterates the official erasure of the human operators and military context that made the kamikaze drone, accepting the proposal that the "electric eye" guides the aerial weapon. Yet, the history of the television-guided assault drone in World War II reveals the "ghostly machine" as a fallible network between human, drone, and kamikaze. These parts are linked and interwoven with the context of the war that is, in turn, shaped by the U.S. experiments.

There is no "drone" that can be separated from the human operator, even if this dissociation is integral to the weapon. The framework for drone aircraft exceeds the technical and human actions that haunt the images onscreen.

Attack! The Drone as Battlefield "Failure"

The ideal drone suggested by *Service Test of an Assault Drone* played out with significantly more complications on the battlefield. Reports from the experimental use of television in 1944 show how the immersive view onscreen was limited. Navy RCA television technologies transferred by Chief of Naval of Operations Ernest King to the army during the summer of 1944 were repurposed to guide "war-weary" B-17 bombers. Through the addition of radio and television control, B-17 airframes were turned into remotely guided weapons. Two top-secret missions, Project Aphrodite and Project Castor, conducted trials with television control in the fall of 1944 in Europe. The operation of "war-weary" B-17s differed from the TDR drones built by the navy. An onboard pilot would take off in the plane from an airfield in England and parachute from the aircraft once the B-17 was airborne. Control would then shift to the remote pilots who guided the aircraft through television and radio control from another aircraft. The pilotless B-17s, laden with twelve thousand pounds of explosives, were flown to strategic targets in the North Sea and in continental Europe, crashing into the site and detonating the explosives onboard. Of the twelve "war-weary" missions that flew B-17s by television control in 1944, only one hit its proposed target, a German oil refinery, undermining the project's ambitions to develop a *guided* missile.

A television-controlled flight, under attack by anti-aircraft fire as it aimed for an airfield in Helgoland on September 3, 1944, was described as follows: "Almost immediately, we could see bursts of flak in the television camera's field of view. In my television screen several miles away at about 5,000 feet altitude, I could see trees, streetcars, automobiles, windows in barracks, an airfield complete with airplanes and the enemy running by the hundred to take cover. . . . The control pilot in the . . . mother plane guided the DRONE, as if he were in it, straight toward the airfield."[41] Using two control planes, the television transmission and radio communication interfaced between the drone and control planes to guide the pilotless B-17 to its target—the German airfield. The drone allowed the television operator both to feel "as if he were in it" and to watch as the plane crashed and exploded while he remained distant and unharmed. Notice that the pilot's description emphasizes the everyday characteristics of the scene below—trees, streetcars, automobiles—before describing a battlefield. Anti-aircraft gunfire onscreen signaled counterattack, registering instead as flickering onscreen. As the image transmission came into focus over the airfield, the picture seems to enable the operator to guide the plane on its trajectory.

Although the pilot's description makes the mission sound as if it were a success, an army summary highlights how the remote-controlled planes in fact failed to locate and hit their targets:

The take-off, set-up route, and bail-out were accomplished without mishap and the two PV-1 control planes guided the robot[42] over the North Sea for the sub pens at Helgoland. The [television] operator, with a poor picture[,] picked up the breakwater on Dune Island, mistaking it for Helgoland. Eight seconds prior to impact, the [television] picture was lost, indicating a hit by flak on the transmitter. The robot also started into a right turn and the controller dove it into Dune Island. The photos show a large crater near the barracks area and extensive damage to the barracks and other buildings.[43]

The contrast between these two reports shows how the supposedly immersive qualities of the television did not ensure that what the remote pilot saw corresponded with what was on the battlefield. While the pilot operated "as if he were" in the television frame, what he saw onscreen did not correspond with the point-of-aim that the guided missile was supposed to attack. The "actual view" is an error.

The army evaluation relies on a particular conception of targeting related to guided missiles. "Success" of the "robot" flight was tied to whether the weapon hit its target. The reports gave no indication of the number of lives taken by the mission, a reminder that the individuals targeted by today's Predator drones correspond to a different framework. Given that the drone damaged the barracks, some casualties seem likely. Yet, the missions are defined by whether they hit their military objective—and only one of the twelve B-17 missions did so. The army concluded, "The results . . . were not satisfactory as far as damage to enemy installations is concerned. However, these missions were in the nature of experimental missions, and have proved the value and serviceability of the weapon and equipment. The failures [though] can be chiefly attributed to weather, with some personnel failure, and in two instances, the possibility of equipment failure."[44] Their report does not substantiate the transparent view between the weapon and the operator claimed for the image transmission system in Zworykin's initial memorandum. Rather, like aerial bombings carried out by manned aircraft, televised air attack was inaccurate and fallible. The "war-weary" B-17s were not successful, strategically. Yet, the army concluded that the missions did prove "the value and serviceability"[45] of the television control system.

The report leaves out not only German deaths but also the U.S. military deaths that corresponded with the two instances of "equipment failure." Two volunteer navy pilots, Joseph Kennedy and Wilford Willy, were killed during earlier tests, after the bombs onboard the aircraft detonated before the pilots who launched the aircraft were able to parachute out of the plane. The back of Hall's book *American Kamikaze* includes a picture of the failed piece of equipment, which apparently caused the plane to explode immediately rather than after a timer went off. Briefly, the editorial decision suggests not the advantage of the American program but its equivalence with the Japanese kamikazes.

STAG-1 carried out other bombing missions between September 27, 1944, and October 26, 1944, in the Japanese-held parts of the Bougainville Islands. The team was split into two groups and the drones were flown in configurations of four, as they were in the test film. The records of the missions provide a careful tabulation of the experiment's progress. Forty-six TDR-1 drones were launched during this month. Of these, twenty-nine assault drones were detonated by their operators, while the others failed before they could be exploded due to mechanical issues, weather conditions, or anti-aircraft fire: in other words, over half were utterly unsuccessful. Of the twenty-nine drones that were detonated, two TDR-1s struck a lighthouse and six hit a beached ship used by the Japanese as an anti-aircraft emplacement. Of the twenty or so remaining, the officials note, "[These] attacks were difficult to evaluate as in most cases the targets were either barely distinguishable or could not be seen at all from the television screen."[46] The report suggests the inconclusiveness of what could be seen through television and the indeterminacy of what happened on the battlefield rather than its definite status. Nonetheless, the navy reports register these missions as hits, and the final analysis of the experiment claims that the weapon hit its target over 50 percent of the time. In addition, Jones and Larkin construed the project as an overall success in their final reports.

Billy Joe Thomas, a remote pilot in the assault drone unit, recalls his experiences flying the television-guided aircraft in the Solomon Islands in an article written in 2005 that linked the STAG-1 mission to the use of Predator drones in Iraq: "Yeah, I got shot down once or twice.... Anti-aircraft fire just brought it down. I didn't have control but the picture was still on the screen, and all of the sudden I was looking straight down and couldn't do anything about it.... If it had been a piloted plane and [I'd have] been shot down, it would have been a funeral."[47] As the remote pilot of the aircraft, Thomas remembers being "shot down." He was not shot down, of course; it was the assault drone he controlled that was hit. His statement explains this separation in the next part of the sentence, recalling how anti-aircraft fire brought "it" down. Thomas describes the picture as out of his control, an image he "couldn't do anything about." Yet he also sees himself through "its" perspective, "looking straight down" as though he sees what the camera sees. Here, the point of view from the camera onboard the aircraft becomes his hypothetical position. In the final sentence, however, he erases his role as the operator, noting that "if" the drone had been a piloted plane, it would have been his funeral.

Hall's third-person memoir *American Kamikaze* offers a structural corollary to the elision between self and object analyzed above. The book recounts Hall's involvement in STAG-1 as if these were not his own memories, thus replicating this distancing between self and object in a slightly different key: Hall distances himself from both the narrative "object" of his memoir and that of his former self, the third-person "he" of the narrative. In what might be another layer of separation between himself and the actions he participated in, Hall reprinted the unit's official correspondence to the navy's Pacific Command in lieu of his own recollection of the military operations between September 27 and October 26, 1944. The title of the book is left hanging, suggesting a broader set of displacements that disavows the American construction of the Japanese stereotype and aims to ameliorate the enmity produced. Yet, the denial is incomplete.

Although the book lauds STAG-1 and the experimental technologies tested by the unit, Hall briefly expresses doubt about his role as a drone operator. According to reports submitted to the Pacific Command, the formal cause of an unsuccessful mission on October 15, 1944 was television failure. Hall explains that, as operator of the aircraft, he knew that the drone actually crashed due to "a partial windup which caused the drone to veer at the last minute and crash almost exactly in the middle of the red cross on the white roof of the hospital."[48] In the navy report, the strike is recorded "at the south end of Hospital Ridge" and makes no mention of a building.[49] The images transmitted through the television persist in Hall's memory forty years later. Writing in the third person, Hall notes, "He couldn't blot out the picture he saw on the [television] screen of the cross looming ever larger and no matter what he did with the stick or rudder controls the drone wouldn't turn, until the screen went blank at the moment of impact." Hall remembers thinking, "What if it really was a hospital, what about all those guys in there, even if they were Japs, what must the survivors, if there were any . . . think of the Americans now after all the atrocities the Americans were accusing the Japs of perpetrating."[50] His control is stymied, and he watches the drone aircraft explode onscreen. The image on the television is one he could not "blot out," even as his position at a distance allowed him to ask "what if" it was a hospital that he struck; he is both part of "the drone" and undone by it.

Cancellation; or, the Drone Is/Is Not Destiny

Robert Jones's final report explains the television-guided weapons could "attack with minimal risk to the pilot and crew."[51] Arguments in favor of unmanned aircraft today rely on a version of Jones's claim: they highlight how the pilot is protected by the distance provided by the networked system and emphasize how technology makes the operator invulnerable. However, the internal navy debate about the use of TDR-1s in World War II demonstrates how the claims of protection and distance provided by networked control are not objective conditions but, rather, part of a wartime context that seeks to objectify the enemy as a target, defined by the exigencies of their time.

Following the month-long test of the TDR-1s, new assignments were issued for the remaining personnel in STAG-1 and "all 30 Avenger [TBM] control planes were placed aboard a barge, taken out to Reynard Sound, and dumped into the lagoon."[52] Commodore Smith laments that the television-guided assault drone program is "dead, buried, and dismembered." Writing to Jones shortly after the unit was disbanded, Smith expresses bitterness. He explains the television-guided drone's failure in terms that emphasize the challenges internal to the navy, while reasserting the inevitability of the technology: "In time of course, the weapon or its counterpart will arise again. . . .It is not an ending for the idea, that will progress in time—to fruition—the making of accurate robot planes and bombs will be solved in 10 or 15 years following the war; instead of being used in this war, as we strived to do. What a source of gratification for those who stopped us."[53] The exchange reflects the determinism that Smith thought drove the weapons project, a view that is immediately undermined by his assessment of the internal resistance to unmanned aircraft within the naval hierarchy.

The narrative of steady, obvious, technology-driven progress remains dominant for Smith, despite its tension with the rest of his account. Opposition to the project was powerful, as comments made by Vannevar Bush, director of the Office of Scientific Research and Development in World War II, showed. In a 1947 letter evaluating the TDR-1, Bush dismisses the project, saying, "We do not need to go into this fiasco in detail. It is an illustration of what can happen when military requirements are written by enthusiasts of little grasp."[54] The strength of this opposition was such that the navy's television-guided weapon was finished even before it had been used in war, and its use in World War II was mostly forgotten.

In 1950, Fahrney retired from the navy and began work on a history of pilotless aircraft and guided missiles. When Fahrney claimed in "The History of Pilotless Aircraft" (1957) that the TDR-1 was the first prototype of a guided missile, his argument was widely discounted. The German use of V-1 missiles, for example, predates the TDR-1 experiments in the Solomon Islands. The putative irrelevance of the assault drone project in the Cold War was not only a question of which project came first but also a question of which became the basis for American missile development. In a series of letters aiming to get permission to publish his manuscript, Fahrney's attempts to portray the drone as a kind of technical manifest destiny ultimately crumble.

In these letters, Fahrney writes to Rear Adm. Paul Stroop, chief of the Bureau of Naval Weapons, while he is in the United States Naval Hospital in

Philadelphia in 1960. The Bureau of Naval Weapons had merged the navy's Bureau of Aeronautics and Bureau of Ordinance two years earlier (the rival bureaus that had once marked the respective positions of Towers and Smith was no longer). In the letter to Stroop, Fahrney reflects on the history of guided missiles he had written in 1957. He explains, "I have had a lot of time to think about many things and one of my recurring thoughts concerns the status of the history of guided missiles." He acknowledges, "It is, of course, abundantly clear to you that I have a very personal interest in the publication of this history because of the part I played in the pioneering effort: but I must submit that I am more concerned with an early presentation of the Navy's distinguished position as the precursor of a new and vastly important era of weaponry."[55] Replying to Fahrney's letter, Stroop writes, "With regard to your question concerning the history of guided missiles, a number of factors have kept us from making progress . . . the Naval Institute after examining your manuscript felt that they were not in position to do the necessary rewriting."[56] A review of the manuscript by Lee Pearson, technical historian for the navy, is more straightforward. He explains, "Admiral Fahrney and I had a frank and friendly discussion of the problems involved in getting his manuscript prepared for publication. . . . Although we never discussed it, I gained the impression that our basic point of divergence is our differing estimates as to the quality of the manuscript he prepared—He considers it needs only an editorial 'cleaning up' while I feel much additional research, evaluation and a complete rewriting is required."[57] Later in the memorandum, in a discussion of the possibility that a "popular history" might be published, Pearson notes, "The manuscript, except insofar as it is classified, is probably in the public domain. . . . (Because of its poor quality, we are trying to classify the manuscript administratively as for 'Official Use Only'.)" It is unclear whether the manuscript was "poor quality" or if Fahrney's account simply did not fit with the ballistic missile program in the Cold War. The efforts to keep the document classified were effective. The National Air and Space Museum has an incomplete copy of the manuscript, which leaves out chapters from the interwar period and beyond. The copy I found in the National Archives belongs to the records of the Bureau of Aeronautics (which includes over 120 boxes of source material that Fahrney also collected); it was not processed and available to the public until 1995.

By 1980, toward the end of Fahrney's life, the U.S. military began to revise its evaluation of the significance of the navy drone project, given the increasing prominence of cruise missiles at the time. Cruise missiles differed from ballistic missiles in that they operated through external guidance controls sent by radio and television, although, until 1987, they were designated by the U.S. military as the same class of systems.[58] In 1980, Fahrney published an article on the television-controlled drone, "The Birth of the Guided Missile," in the *United States Naval Institute Proceedings*,[59] followed by another article, "The

Genesis of the Cruise Missile," in *Astronautics and Aeronautics* in 1982. His piece in *Astronautics and Aeronautics* argues, "The cruise missile, in form and substance, was evolved and creditably functioning in the closing months of WWII as a part of the pilotless-aircraft program carried out by the Navy's Bureau of Aeronautics."[60] These articles' publication reflects the newfound interest in guidance systems in the 1980s. The drone returns in a changing global context, which calls for a new story of targets and targeting.

3

Unmanning

● ●

In the Cold War, the U.S. military tested the drone for a new experimental use: unmanned surveillance. A classified position paper for the air force and Department of Defense written in 1961, "Alternative Reconnaissance Systems," advocated for this use. "Alternative Reconnaissance Systems" was circulated by the military after Francis Gary Powers, an American U2 pilot, was taken prisoner while flying a secret aerial reconnaissance mission. The crisis that ensued following Powers's capture heightened political tensions between the United States and the Soviet Union, though the incident took the U.S. public (and Congress) by surprise, as both were unaware of the classified reconnaissance missions with which Powers was tasked. The report proposes that future overflights for aerial reconnaissance would be "unmanned—for political, diplomatic and public acceptability; decreased detectability due to size; decreased design sophistication; increased operational flexibility; increased security and cover."[1] The report positions "unmanned" flight as a counterpoint to political, diplomatic, and public repute, indicating that the "alternative" system would provide a "technological" solution for global surveillance. A year after the report, aerial reconnaissance would be used in the Cuban Missile Crisis to provide evidence of Soviet missile buildup and, later, withdrawal. No drones were deployed. The retrieval of images from aerial flights, however, was invoked by the government as a national security measure for the U.S. public. A view from above both set out an objective, technologically given picture and acted as a response to new form of threat that emerged in the Cold War.

This chapter considers the rise of this new term "unmanned," which uses the prefix *un-* to oppose the drone system to manned flight. The disavowal proposed by the name organizes politics and technology as separate fields in the early stages of the Cold War, even as the term never undoes human action. In World War II, drones were described as targets and prototype guided missiles, and their flight was "pilotless" or "manless." By 1946, usage shifted to "unmanned," indicating the negation of man, not just his removal, even as drones proliferate as pilotless target aircraft. The denial enacted by "unmanning" positions drone flight in opposition to, rather than as an extension of, piloted flight, while the drone's eventual use for surveillance by the United States disavows human action to outline its function as apolitical. The drone is constructed as monitoring the nation rather than enacting its politics.

A precursor of the conceptual shift from pilotless to unmanned aircraft is suggested in the final stages of World War II in a public account of the secret assault drone program. A *Washington Daily News* article from February 12, 1945 offered a description of the military's drone use in Europe (related in chapter 2). It declares the pilotless aircraft the " 'Brain Box' of the Air Forces" and recounts how "war-weary" B-17 bombers were repurposed for aerial assault by remote operators. The article describes the weapon as a "brain box," suggesting that its control systems were based on internal mechanisms. Operators would have continued to be necessary for flight, though, and their role is unmentioned in the article. Instead, the report names a "robot brain," suggesting the missions are the result of semi-autonomous flight. "Instead of being scrapped, these veterans of many raids are filled with explosives, taken aloft by their crews, who set the automatic pilots to fly the planes against German targets, and then bail out. The big Fortresses, like work horses headed for the barn, do the rest themselves, guided only by the robot brain in the automatic pilot."[2] The article positions B-17s, like the target drones before them, as human "veterans" and as an anthropomorphized machine that will automatically defend the nation like obedient "work horses on the way to the barn." Meanwhile, the action of the crew is passive; they are merely passengers "taken aloft by the plane." Once the crew bails out, the aircraft "do the rest themselves," acting through "the robot brain in the automatic pilot." In the description, the role of the system shifts from being aligned with a soldier (the "veteran") to an obedient animal once the "robot brain" has taken hold (the "work horses"). The article omits any mention of the failures of the supposed innovation as well as the eventual fact that the project would be discontinued in favor of the ballistic missile systems that were the focus of Cold War military weapons development.

Although the project to use drones as guided weapons was cancelled, unmanned aircraft continued to be a platform for training air defense in the Cold War (similar to programs discussed in chapter 1), where the "drone" assumed a public role as an enemy target. The jet-powered systems were launched from a catapult and landed by parachute, avoiding the challenges faced in the earlier programs to achieve pilotless takeoff and landing. The aircraft could be preprogrammed to fly on particular flight path; however, remote pilots on the ground and in control aircraft continuously monitored the drones. These aircraft were termed "unmanned," and although they continued to network together human, machine, and media, public accounts relied on anthropomorphizing descriptions of machine autonomy. Articles on drone aircraft in aeronautical journals in the 1950s drew on the language of cybernetics, which studied systems as feedback, outlining the function of radio control and programmed guidance used to operate the drone as an "electronic brain."

Secretly, manufacturers modified target drones for unmanned surveillance and promoted them as an alternative to manned reconnaissance beginning in 1960. In the classified "Alternative Reconnaissance Systems" report, the claim was advanced that unmanned aircraft would "assist [the U.S. military] in gaining Executive approval since the political risk is minimized due to the absence of a possible prisoner."[3] In this way, the discourse of machinelike autonomy that circulated in the public was tied to a position within the government that promoted unmanning as a technological alternative to political risk. Unmanning in the Cold War linked machinelike feedback with a cybernetic context for global control and aerial surveillance. It disavowed politics in two ways: first, by proposing that enmity and protection could be defined automatically and, second, by in turn disavowing the military, government, and industry decisions that shape these practices. In the United States, the period was marked by the simultaneous rise of the military-industrial-academic complex.

Paul Edwards, analyzing Cold War–era computation, describes the overlay of cybernetic language, technology, and global military control through the term "closed world." Unmanning fits within this broader discursive shift. He writes, "The phrase 'closed-world discourse' describe[s] the languages, technologies, and practices that together supported the visions of centrally controlled, automated global power at the heart of American Cold War politics."[4] One ideal of centrally controlled automation, key to Edwards's analysis, emerged from the Semi-Automatic Ground Environment (SAGE) Air Defense System, built at MIT's Lincoln Laboratories. It aimed to partially automate air defenses against atomic weapons. SAGE responded to a newly defined global sphere marked by atomic attack, which emerged after the Soviet Union first tested nuclear weapons in 1949. The potential for global war through nuclear bombardment extended national defense beyond the territory of the United States to its skies, a "closed world." Domestically, public accounts of these projects in the United States came to portray national defense as an aerial project, tying an automated sphere of global control to a citizen-subject prepared to respond to threat automatically.

Unmanned aircraft are just one of countless developments in the Cold War marked by the twinned impetuses of automated control and national defense; indeed, their significance and use in this period were ultimately inconsequential in comparison to nuclear weapons, computation, and satellites. Target drones were reserved for simulations of air defense, while unmanned reconnaissance aircraft, based on the target system, were deployed for a few thousand flights in the Cold War, mostly in China and Southeast Asia. By 1980, all of the U.S. military's projects to use unmanned aircraft for surveillance were cancelled. Drones were neither fully automatic nor all-seeing. Yet, they lay claim to machinelike autonomy nonetheless. Edwards notes how closed-world discourse exceeded the programs' capacities, marking their misalignment with the conditions they aimed to control and their use by the U.S. military.[5] "Unmanning" does not just describe the removal of the pilot's body from the plane, then, but an affective dimension of U.S. national security that disavows human action. Drones' failures highlight how discursive claims about the power of unmanning are undone in practice. A machinelike context of global control is unachievable.

Explaining what he calls the "national security affect," Joseph Masco writes of the Cold War United States, "The goal of the national security system is to produce a citizen-subject who responds to officially designated signs of danger automatically; instinctively activating logics and actions learned over time through drills and media indoctrination."[6] Unmanning is a counterpart to these practices; it establishes a framework for an automatic citizen-subject by producing notions of a machinelike world. The U.S. government, as an emerging superpower in the Cold War, proposed to act globally through a set of measures conceived as distinct from political, diplomatic, and public spheres. Yet, unmanning is not apolitical. As a concept, unmanning normalizes the myth that technology, or indeed any human pursuit, exists outside of and without politics. The limits to drone flight during the Cold War underscore how the United States' technocratic claim to automation and objectivity remained invested with ghostly ties to human actions and institutions. The national automation of the citizen-subject in Masco's analysis is simultaneously made by the haunting that denies human action, which remains central to these projects nonetheless.

A new global context, defined by the possibility for atomic war, was a central concern for Hannah Arendt. In *The Human Condition*, originally published in 1958, she outlines how automation conflates conditions set into motion by human work within the natural world. She traces the use of tools used to modify and shape the world to the rise of mechanization, in which "the world of machines has become a substitute for the real world." For Arendt, technology—instead of establishing means and ends—instead reshapes the world as a machine. She writes, "The question . . . is not so much whether we are masters or slaves of our machines, but whether machines still serve the world and its things, or if, on the contrary, they and the automatic motion of their processes have begun to rule and even destroy world and things."[7] However, Arendt's concern with the rising significance of automation in the Cold War might be ameliorated by engaging with its failures.

Reading unmanning as a disavowal troubles accounts of a straightforward rise of automation. Arendt writes, "The discussion of the whole problem of technology . . . has been strangely led astray through an all-too-exclusive concentration upon the service or disservice the machines render to men." In these statements, Arendt ties the postwar globe to a technology that can "rule and destroy" its makers. The destruction made possible by the atomic bomb illustrates for Arendt "the enormous scale on which such a change might take place."[8] In a world defined by massive destruction, the automaticity of machines becomes interwoven not only with labor but also with government. Rather than a common, shared world made through politics, this world's terms are laid out instead as technical—a world simultaneously objectively given and more-than-human in its capacity for destruction. While the narrative of unmanning revealed across this book was conceived as replacing human action, this same history read closely shows that the mechanized world Arendt fears never fully takes form: it is always a human hand that drops the bomb. The awesome potential for destruction that arose in the Cold War cannot be minimized by the image of a few crashing drones. Yet, those drone crashes highlight how the affective and material dimensions of the Cold War never fully created the machine totalities they claimed. Rather, human actions were made spectral and disavowed.

The U.S. government, military industries, and individuals affirmed an opposition between politics and technology during the Cold War. This claim to a technology wiped clean of politics is made by the U.S. military-industry for unmanned aircraft after the capture of Francis Gary Powers: unmanning mitigates political risk. Yet, unmanning only tentatively takes hold, complicating the purportedly smooth evolution from manned to unmanned that engineers maintained to be the aircraft's basis. To this day, despite decades of advancements with remote sensing, including now ubiquitous unmanned aircraft and satellites, manned reconnaissance aircraft such as the Lockheed Martin U-2 or "Dragon Lady" are still in service; manned and unmanned systems continue to operate in tandem, interlinking human and nonhuman. Arendt might use this example to point out how human action has become indistinguishable from technology and the prevailing logic of a machinelike world. But I argue that the reverse is in fact the case: unmanning is a contingent set of situated practices imperfectly made by human, media, and machine. If machines do not produce a world that is given but one that is instead indeterminate and flexible, entangled with human actions rather than distinct from them, technology is not the opposite of politics but rather a focal point for politics. Thus, it is not automaticity that seems to foreclose the possibility of creating a groundwork for politics but the paradoxical denial of human action through technology.

"The Bee with the Electronic Brain": Cybernetics as Destiny

At the end of World War II, drone targets were mass-produced for anti-aircraft trainings. These were miniaturized versions of the aircraft developed by the navy beginning in 1936, which had their origins in a model airplane shop in Hollywood run by the actor Reginald Denny. He sold his company, Radio-plane, to Northrop in 1950. Its chief rival for the production of drones in the postwar period was Ryan Aeronautical, which made a drone called the Fire-bee. Ryan Aeronautical won a competitive contract from the Department of Defense to produce jet-powered target planes in 1948. The Firebee mimicked attack aircraft by drawing on the increased speed of jet planes. The targets were used to train surface-to-air and air-to-air defenses, replacing drone targets built during World War II, now too slow to accurately replicate aerial attack in the postwar period. The jet-powered target marked a significant shift for Ryan Aeronautical, which had been known for its pilot training aircraft and pilot training program during World War II. All three branches of the military used Firebee drones to train for air defense throughout the Cold War. Northrop eventually gained a monopoly over the target planes, acquiring the Firebee system in 1989. Firebee drones continue to be used for training today. As a simulation, the target plane created conditions of war to which increasingly automated systems of attack and defense were to respond.

Beginning in 1953, Ryan Aeronautical developed a marketing campaign for its drones through the company's magazine, press releases, and a number of articles that appeared in aeronautical magazines. The promotional materials utilized cybernetic imagery to position the target plane as a critical innovation, despite the fact that it represents a minor investment for the U.S. military in comparison to the jets it mimics or the missiles tested against it. Ryan Aeronautical lauded the Firebee in an article "The Bee with an Electronic Brain," published March 15, 1953 in the company's magazine, *Ryan Reporter*. With the bravado of self-promotion, Ryan announced that the once-secret project would now be public, crowing, "the spectacular Ryan 'Firebee,' from which the curtain of secrecy was recently lifted by the Department of Defense, is America's newest turbo-jet, pilotless target drone, capable of near sonic speeds at high altitudes." The subheading outlined its function for the United States, noting "Ryan's Fire-bee, America's newest turbo-jet pilotless target, duplicates performance quali-ties of jet planes in combat over Korea."[9] The Firebee target mimicked enemy aircraft built to fly faster than the speed of sound but demonstrated that the United States also had this capacity. Its "spectacular" use is tied to U.S. military performance, which sets up defense and attack through the figure of the drone, promoting a machinelike context of war directed by an electronic brain.

The Firebee is not merely a simulation of *piloted* aircraft flown by enemies; it also indicates changes to *pilotless* aircraft. Human pilots never tested the Firebee. From its beginnings, the system was designed and engineered to be unmanned. This new sense of the Firebee is captured by the title "The Bee with an Electronic Brain." Not only is the drone aircraft likened to a bee—similar to the drone targets discussed in chapter 1—but the "electronic brain" suggests that the machine is capable of generating its own responses, likening the operation of the system to the feedback loops that define cybernetic theories that were also prominent at the time. Cybernetics was a term invented by Norbert Wiener in 1948 to describe a multidisciplinary approach to the study of control and communication, from the Greek for "steersman."[10] The reach of cybernetics during the Cold War was wide-ranging, influencing biology, ecology, computer science, communications studies, social sciences, and the arts. It is the ways that these cybernetic theories became part of language used to describe the Firebee in 1953 that I attend to here. It is unclear to what extent cybernetics informed the engineering decisions made by Ryan Aeronautical during the process of building the Firebee between 1948 and 1953, if at all. However, the press releases rely on the ubiquity of the ideas, describing the drone as having an "electronic brain" and a "black box" operation to describe its unmanned function.

"Behavior, Purpose, and Teleology," a foundational text for cybernetics written in 1943, distinguishes between purposeful and purposeless behaviors in order to transform the definition of machines: rather than mere tools, technical systems, like organisms, can act in response to feedback. To draw out this distinction, the text relies on the imagery of a mechanical attack. The authors, Arturo Rosenblueth, Norbert Wiener, and Julian Bigelow, suggest that "a torpedo with a heat-seeking mechanism" might be "intrinsically purposeful," as its response is always guided by its reaction to heat.[11] Cybernetics, in this way, situates action relationally, occurring between the object's aim and its environment. The purpose of the object is linked to its reactions. To align organisms and technologies, cybernetics conceives of both as systems organized through inputs and outputs. This ideal—a standard, monadic unit governed by feedback—elides the relation between human and machine, however. It distinguishes one from the other even as their "purposeful" reactions are connected.

The Firebee drone's operation is not autonomous, though its flight appears purposeful. The Ryan Aeronautical article describes its function as follows: "Responding to ghost like controls that may be miles away, Ryan Firebee flashes across the sky, ready to simulate fighter plane tactics in sharpening anti-aircraft defenses."[12] In the passage, a passive controller—presumably a human being— is figured as a phantasm, motivating the response of the Firebee flashing across the sky. The reader is invited to see the drone *as if* it responded to high-speed attack directly. Its flight is presented as an automatic response to global threats that the United States faced in Korea and elsewhere. The machine's function is activated by ghostlike controls, while being presented as a uniquely American achievement. Through cybernetics, Katherine Hayles examines how the body is reconfigured as an information system, destabilizing the ontological foundations of what counts as human by analogizing organism and machine as autonomous, purpose-driven systems.[13] Similarly, the "electronic brain" in the account promoted by Ryan Aeronautical uses the ideal of a brain to describe the action of the drone. This description denies the mediations that link operator and aircraft—the actual basis for the system's function—suggesting instead that the drone acts on its own impulse.

"A Bee with an Electronic Brain" goes on to explain that the drone uses a "small black box containing a control stick and switches to govern engine speed and other flight conditions, and to transmit control signals to the drone."[14] A black box organizes relations between human and machine as a single, behavioristic unit. The box is described as sending electronic transmissions to the Firebee, governing its actions. The article notes, "By use of the ground remote control station, the 'nolo' (no live operator) aircraft can be flown out-of-sight at high altitudes, while other men on the ground track it by electronic devices."[15] "Nolo" indicated a solo, machine flight, a term used in trials with drone targets in World War II. It is replaced entirely by the term "unmanned" by the 1960s. The Firebee target is disembodied and immanent when it flies out of sight, flashing across the sky, while it appears both embodied and distant when operated by men on the ground. It is the interplay between these positions that characterizes the machinelike function of the Firebee and the spectral presence of the operator. When characterized through a framework of inputs and outputs, the Firebee confuses the question of who or what responded to the external conditions. The target's response, organized by human operators and preprogrammed flight patterns, is made to seem self-determined and purposeful through its "black box."

The 1953 magazine article provides few technical specifications, which are found instead in a declassified presentation, "Firebee I—A Case Study in Pilotless Aircraft Evolution," released with permission of the air force in 1981.[16] "Firebee I" explains that the original Q-2 flight control system, developed after Ryan was awarded the defense contract in 1948, responded to five radio control commands: (1) climb, (2) dive, (3) right and left turns, (4) straight and level, and (5) engine rpm increase-decrease to control airspeed.[17] With only five commands, the radio controls were simpler than systems developed by the navy during the interwar period that used a telephone dial. The jet-powered drone flew faster and farther than previous systems, beyond the range of human sight, and was therefore to be tracked electronically. In the early test flights, the controller easily stalled the drone (which would then crash) because there were no visual cues for operating the aircraft. "Firebee I" explains how engineers later preprogrammed the engine power and speed of the drone to correspond to certain climb and dive rates, which made it more difficult to stall. While the control box would send the signals to maneuver the aircraft, the speed and pitch associated with these changes would rely on programmed responses—that is, an "electronic brain."

In the article "The Bee with an Electronic Brain," the "black box" confuses the operator's control and its internal, preprogrammed function, as if this were a single, automatic unit. The author comments, "[The] push button heart of the Firebee project is a small 'black box' containing a control stick and switches to govern engine speed and other flight conditions, and to transmit control signals to the drone."[18] The elision of human engineering, design, and control with a behavioristic model of technology provides the conditions for "unmanning" to emerge, as if the black box exceeds human action. "Black box" typically indicates a closed system, especially in cybernetics. The definition in the *Oxford English Dictionary*, however, includes an example from an *Aeronautics* article from 1932: "For the sending of control messages, there is located on the destroyer a little black box." The example sentence locates the box's use as a message center, not a closed control system.[19]

Donald MacKenzie sets up the centrality of the black box to science and technology studies through a quote by Charles Draper, founder and director of MIT's Instrumentation Laboratory. Draper explains that the black box is an ideal arrangement of missile guidance: a self-contained unit that would not be affected by external conditions: a closed system. For MacKenzie, this leads to a broader meaning: "It is a technical artifact—or, more loosely, any process or program—that is regarded as just performing its function without any need for, or perhaps any possibility of, awareness of its internal workings on the part of users."[20] MacKenzie comes to this definition by showing how the guidance system troubles Draper's idea of an apparently self-contained system. MacKenzie argues, to the contrary, that guidance technologies are inextricable from the conditions that make their development possible. Extrapolating from this analysis, MacKenzie writes that "the more deeply one looks inside the black box, the more one realizes that 'the technical' is no clear-cut and simple world of facts isolated from politics."[21]

Inside the Firebee's black box there is an overlay between remote communication and a self-regulating function. The elision of one for the other itself yields a context that disavows politics, claiming that the Firebee's flight is driven by an internal, mechanical purpose. It is not just that technical systems are shaped by political, social, and economic processes but also that the simultaneous confusion of the former for the latter defines national security as inherent to the Firebee's function. In another image and caption, also from 1953, a Ryan Aeronautical press release describes the drone (figure 9): "Like a

FIG. 9 Firebee drone, launch plane, and press caption, Holloman Air Development Center, NM, 1953
Credit: National Air and Space Museum

released parasite, the Ryan Q-2 pilotless drone target plane is launched from its B-26 'mother' plane and streaks out over the desert under its own power during U.S. Air Force development tests at Holloman Air Development Center, Alamogordo, N.M. Speed and maneuverability of the 'Firebee' are controlled from the ground by means of a black box remote control which transmits command signals to its electronic 'brain.'"[22] The caption initially presents the drone as a "released parasite," launched from a "mother" plane. The Firebee is made foreign on its release, turned into the enemy target it is supposed to mimic by emphasizing its parasitic qualities: a parasite is simply another term for a foreign body. Turning to the system's "speed and maneuverability," however, the tone changes. The parasitic foreignness becomes a black box remote control that transmits "command signals" to the aircraft. In this latter part, the qualities typically associated with automatic control take precedence, while the role of operator disappears into the black box used to guide the aircraft.

Automaticity is tied to the drone's being made strange, linking to enmity and division not just a set of technical conditions but also a disavowal of human action. Discussing the rise of cybernetics, Peter Galison outlines how Norbert Wiener's attempts to build an anti-aircraft predictor during World War II inform its premises, setting out a "cold-blooded machinelike opponent" as a basis for feedback.[23] Cybernetics shapes "an image of human relations thoroughly grounded in the design and manufacture of wartime servomechanisms and extended, in the ultimate generalization, to a universe of black-box monads."[24] Through cybernetic theory's origins in an anti-aircraft predictor, Galison proposes a new ontology of enmity in the Cold War, based on a mechanical exchange between opponents. He distinguishes this new enemy from two other versions. The first is a barely human Enemy Other often associated with "lice, ants, or vermin to be eradicated" and aligned with the Japanese. The second enemy, the target, is "an anonymous object of air raids.[25] The Firebee, however, complicates this distinction, as the machinelike opponent remains both enmeshed with a parasitic other and at an anonymous distance. Rather than achieve automaticity, the Firebee equivocates between human and machine, layering the automatic conditions of war in the Cold War with the enemy-as-other and target, negating these qualities as an "electronic brain" that is never entirely machinelike.

Open Skies: Making Strategic Reconnaissance Fact

In the Cold War, the ideal of strategic reconnaissance, promoted in a 1951 booklet published by the Rand Corporation, was the ongoing, worldwide collection of images as if the entire world could be pictured and, thereby, known. Global reconnaissance was part of the context of automated war created by jet aircraft, missiles, and automatic weapons. The emergence of strategic reconnaissance marks a shift from earlier aerial reconnaissance, collected largely during wartime and conceived within and as a part of military strategy. Instead, surveillance becomes part of peacetime mobilization tied to an atmosphere of continuous threat. Its practice was connected to two new military institutions inaugurated in the postwar period: the Central Intelligence Agency (CIA) and the air force. While the automatic functions of drone targets were promoted to the public as a mechanized "enemy," unmanned aircraft were secretly tested by the United States as a platform to collect surveillance. Unmanning links the ideal of machine control to an objective picture of the world, defined by an aerial image. Disconnected from the human actions and institutions that produce reconnaissance, aerial pictures were understood to be self-evident images. Both of these aims fit with the goal of so-called national security in the Cold War United States, which itself was tied to the continuous threat of automated war. National protection would take the form of a flying camera. The camera unmans the actions of operators, engineers, government officials, industry executives, and others who make the systems function while defining threat and attack in the Cold War as a system that operated automatically and without human volition.

Prior to the Cold War, aerial photography was tied to military strategies to monitor troop movement, locate targets of aerial bombardment, and assess operations; beyond war, aerial images were used for mapping and urban development.[26] These functions were distinct from spying. During a conference held by the Royal Air Force, the Royal Canadian Air Force, and the U.S. Air Force in 1948, the idea of continuous aerial surveillance was discussed as a strategy to counter the rising power of the Soviet Union. A key proponent was former reconnaissance-airman-turned-Kodak-sales-representative Lt. Col. Richard Leghorn. He explained, "Long-range strategic reconnaissance should be employed today as a means of peacetime spying against the Stalinist empire. . . . We are essentially at war even though this war today is limited, and we must have information on the Russian military and industrial system and capabilities, together with knowledge of Russian intent."[27] Leghorn promotes aerial surveillance as a way to monitor the enemy in a "peacetime" war, waged through military and industry buildup. He warns, "Today the Russians can block to a large extent all our techniques for gathering information, except military aerial reconnaissance." Leghorn advocates for the use of stealth aircraft and unmanned systems because "any aerial reconnaissance we conduct over Russia today must be extremely difficult or even impossible to detect." Within the newly formed air force, Leghorn's proposals initially gained little traction, as the primary concerns for the military at the time were jet-powered aircraft, nuclear weapons, and missile technologies. They would, however, be eventually taken up by the also recently formed CIA.

In the United States, automated global control extended the function of national security to scientific research and industry. After the Soviet Union tested the hydrogen bomb in 1953, President Dwight D. Eisenhower sought answers from scientists, not just military advisers, to examine how U.S. vulnerabilities to Soviet attack should be addressed. In 1954, he commissioned James Killian, president of MIT, to lead a Technological Capabilities Panel. In his final report, Killian explained that the panel was "a technical task force to study ways of avoiding surprise attack by a searching review of weapons and intelligence technology."[28] He emphasized how scientific advancements would serve the U.S. state, foregrounding a model of national defense based on technocratic measures. The still partially classified report focused on whether the United States would be able to detect an aerial attack from the Soviet Union, concluding, "There is a real possibility that a surprise attack might strike without useful, strategic early warning."[29] To counter these vulnerabilities, the panel report provides recommendations to the U.S. government to maintain the offensive advantage that the United States claimed over the Soviet Union by promoting missile development.

The panel's emphasis on technology extended beyond the buildup of arms. Killian selected Edwin Land, the founder of Polaroid, to lead the subcommittee on intelligence for the Technological Capabilities Panel. The subsection devoted to the topic of intelligence observes: "We *must* find ways to increase the number of hard facts upon which our intelligence estimates are based."[30] The appeal in the report to hard facts simultaneously aimed to identify how such intelligence could be produced. Facts will be made by technology. As an answer to this call, the report claims "revolutionary new techniques will be devised to give us facts and answers instead of assumptions and estimates."[31] Image collection was central to this strategy, reworking earlier versions of military reconnaissance as instead a strategy of peacetime national defense. Sections of the report remain classified to this day, including six paragraphs (about a half-page in length), outlining key recommendations for the collection of intelligence, and an entire section titled, "Intelligence: Our First Defense against Surprise."

Kristie Macrakis uses the term *technophilia* to describe the love affair American intelligence has with technology. This form of technophilia differed substantially from the methods used by the Soviet bloc, which relied on human intelligence cultivated through a network of agents to collect information, which proved more successful. Despite the array of technologies created for U.S. spy programs, Macrakis contends, "the Eastern Bloc won the spy wars because of a more effective espionage style." She observes that the Eastern Bloc's great success was in using double agents, noting, "At the end of the Cold War, the CIA discovered that all of its East German and Cuban agents were, in fact, double agents working at the behest of East German or Cuban foreign intelligence." The advantage of human agents, Macrakis explains, is that they could "provide [a] discovery with context and meaning; machines said nothing about the *intentions* of Soviet leaders." Macrakis writes that in the Cold War United States, however, "technology had become a cure-all, a 'fix' for numerous problems."[32] U.S. technophilia was supported by the rise of the military-industrial-academic complex, to which the development of the "revolutionary new techniques" advocated for in the panel was contracted.

Although the precise proposals are still secret, Land and Killian were instrumental in securing Eisenhower's approval to build the U-2, a stealth reconnaissance aircraft, and the report is credited with solidifying the importance of strategic peacetime reconnaissance.[33] The U-2 was developed through the CIA and built at Lockheed's Skunkworks. Land probably came across Lockheed's proposed U-2 plane (which had been rejected by the air force earlier) while reviewing materials for the Technological Capability Panel report. Land and Killian proposed that the U-2 be flown by the CIA rather than the military, which they argued would help buttress the claim that it had a peacetime, rather than military, function.[34] Between 1956 and 1960, the U-2 was flown twenty-four times over the interior of the Soviet Union in top-secret missions known only to a small group of air force and CIA personnel, scientists, and the president. Only four members of Congress were briefed on the flights, which remained secret until 1960.

With the U-2 project, the secret worldwide collection of aerial and electronic intelligence became part of U.S. Cold War strategy. Collection of aerial photography was also the basis for Eisenhower's "Open Skies" proposal for mutual surveillance between the United States and the Soviet Union. It was rejected at diplomatic meetings held in 1955 between the two countries as an intrusion of sovereignty, though the U-2 project remained classified and continued to develop. The government's increased investments in strategic reconnaissance reflected a larger technoscientific strategy for the United States. That strategy relied on the veneer of scientism for its construction of "hard facts," despite those facts often being faulty, partial, or simply incorrect. The refusal of the "Open Skies" policy in the 1950s, moreover, points to the militarism that remains implicit in global reconnaissance despite its claims to neutrality and objectivity, as well as the differences between U.S. and Soviet intelligence strategies. Aerial reconnaissance images accrue the qualities of neutrality and fact when the practices that produce those images are removed from the picture itself. The U.S. brand of technophilia relied on the ideal of a closed world not just for cybernetic systems but also as a means to control information. Paul Edwards writes, "Containment, with its image of an enclosed space surrounded and sealed by American power, was the central metaphor of closed-world discourse. Though multifaceted and frequently paradoxical, the many articulations of the metaphor involved . . . far-reaching military commitments that entailed equally far-reaching domestic policies."[35] In this way, unmanning is significant not only as an intelligence strategy in the Cold War but also for the ways in which it translated the technological solutions provided by the military-industrial-academic complex to the U.S. public.

When the U-2 was first flown over the Soviet Union in 1956, ground-to-air defenses were unable to reach the aircraft, which flew out of the range of current missiles, even though the flights could be detected by radar. The flights were effective in collecting photographs of the Soviet Union, though not in remaining "secret" per se, as they were detected by the Soviet military.[36] After the launch of *Sputnik* in 1957, Soviet missile capabilities improved. In official reports issued in March 1960, the air force and CIA reiterated concerns about the susceptibility of U-2 reconnaissance to anti-aircraft missiles. These reports circulated among insiders involved in the project, three months before Francis Gary Powers's capture.[37] Both the CIA and the air force concluded that the latest Soviet surface-to-air missile could intercept a U-2 mission. Nonetheless, the CIA continued flights.

It was in this context that the air force proposed the use of drones as an alternative. Lt. Col. Lloyd Ryan, deputy to the chief of the Reconnaissance Division at the headquarters of the U.S. Air Force, recalled being in the basement of the Pentagon in 1959: "We were discussing what we would do if and when a U-2 was shot down."[38] Ryan considered the matter with his commander, Col. Harold

Wood, over the next several days, and a possible solution was proposed during a visit by Ray Ballweg, vice-president of Hycon, the manufacturer of the camera system aboard the U-2. Ballweg suggested the air force use drones for reconnaissance. Ryan recalls, "Our response, Hal Wood's and mine, was, 'what drone?' We didn't know anything about drones."[39] "Unmanning" was not an obvious solution to those in power but, rather, a heretofore unconsidered possibility.

At the time, Ryan Aeronautical supplied jet-powered target drones to the U.S. military. The system was not yet a reconnaissance platform; rather, it was a high-speed target with a so-called "electronic brain." The following year, in 1960, Ryan placed Robert Schwanhausser in charge of its reconnaissance drone program.[40] Schwanhausser was hired by the company after serving as an air force project engineer for the Firebee program at Holloman Air Force Base in New Mexico between 1952 and 1954. Schwanhausser and a team of nine engineers set out plans for a reconnaissance system based on the Firebee over the next three months. With coaching provided by Ryan and Ballweg, Schwanhausser briefed the Air Force Reconnaissance Panel at the Pentagon on April 21, 1960, arguing, "The use of U-2 manned vehicles for overflights of the territory of nations unfriendly to the United States creates, we believe, risks which are unnecessary to take. We feel there is a solution to this in the logical evolution of the unmanned Firebee drone system."[41]

On May 1, 1960, just two weeks after Schwanhausser's briefing, CIA pilot Francis Gary Powers took off from a secret U.S. air base in Peshawar, Pakistan. Soviet radar tracked the plane over Afghanistan, and the U-2 was shot down while flying over Sverdolvsk, Russia. Powers was able to bail out of the U-2 by parachute and was captured, along with what remained of the plane. When the U-2 failed to arrive at the base in Norway, however, it was unclear to the Americans what had happened. Initially, the White House tried to cover up the spy mission by claiming the U-2 was a weather plane that had flown off course. Soviet Premier Nikita Khrushchev exposed the falsity of the U.S. story by announcing the U-2 pilot was still alive. Powers's plight became a central narrative to the Cold War, and his story circulated widely in U.S. media and abroad. The incident came just two weeks before Khrushchev was to meet Eisenhower in Paris, along with the leaders of France and England. On the morning of the summit meeting, May 16, 1960, Khrushchev described the spy flights as a "provocative act," and he left before the talks began. He declared, "We are unable to work at the conference . . . because we see from what position it is desired to talk to us—under threat of aggressive intelligence flights."[42] The speech underscores the contradictions of U.S. reconnaissance: the United States pursued so-called peacetime intelligence collection through a logic that suggested the country was "essentially at war."

Powers's capture did not lead to a reappraisal of the reconnaissance strategy, though Eisenhower did discontinue U-2 flights over the Soviet Union.

Rather, *the pilot* was highlighted as a weakness, and the pursuit of reconnaissance methods that were conceived of as wholly technological gained traction. In the next decade, photographs from satellites and drone aircraft, in addition to manned overflights, became standard methods for collecting aerial reconnaissance. After Powers was captured, Schwanhausser recalled, "Things started to happen very rapidly."[43] Yet, the logical evolution of drone reconnaissance that Schwanhausser foregrounded in his presentation to the air force was not as straightforward as he made it seem. Following the Powers incident, Ryan Aeronautical was given its first exploratory contract during the summer of 1960 for a project known as "Red Wagon" within the air force. In a letter from August 19, 1960, T. Claude Ryan, the president of Ryan Aeronautical, wrote to Dr. Joseph Charyk, the undersecretary of the air force, outlining how the company would manage the program as well as the facilities and working capital available for the project. The contract that would have allowed the project to continue was not approved by the secretary of defense, however, who returned the proposal with a brusque note: "I thought we weren't going in this direction."[44] Although Powers's capture by the Soviet Union in 1960 provided a rationale for the project, unmanned aircraft contracts continued to compete with manned reconnaissance aircraft and satellite contracts. The drone's utility, then, was not a foregone conclusion. Eventually, on February 2, 1962, the air force awarded a major 1.1 million-dollar contract to Ryan Aeronautical for four Q-2C Special Purpose Aircraft (SPA), which would be modified for photo reconnaissance and developed to evade detection by radar.[45] Beginning with this contract, Ryan Aeronautical would produce twenty-nine versions of the Firebee reconnaissance system over the next thirteen years (these systems are discussed in chapter 4). However, their scope was never as far-reaching as imagined, and, ultimately, they never flew over the Soviet Union.

The analysis the company provides of the Firebee in 1981 reiterates the evolution described by Schwanhausser. The authors frame transformations to the Firebee based on adaptation to new and changing conditions: "The element which, more than any other factor, allowed the Firebee program to survive for such a long period of use was the ability to adapt to new and changing conditions. The total Firebee evolution was not planned. In fact, the growth of the Firebee targets program was like 'Topsy'—it just grew."[46] The authors refer to the survival of the drone through a striking figure of speech describing something that grows without external design, appearing to increase by itself. The phrase "Grow'd like Topsy" is from Harriet Beecher Stowe's book *Uncle Tom's Cabin*. In the book, the phrase outlines a racialized depravity ascribed to Topsy for the presumably white, Christian reader. The phrase, then, must be situated in relation to these origins, which complicate its use as a description of the drone's technological evolution and suggest instead how notions of unmanning draw on racist cultural stereotypes. The claim made in the documents emphasizes the ability of the machine to respond and develop on its own, even as the description simultaneously undoes this idea by tying the technology of unmanning to a racially circumscribed ideal of social evolution.

Even though the evolution proposed by Ryan Aeronautical is distinct from the cybernetic language of the black box and the electronic brain in the 1953 press release, these accounts emphasize the ability of the machine to respond and develop on its own. The technical report leads the reader to believe that the drone made itself, suggesting cybernetic—and a racialized dystopian—machine autonomy. The Firebee, however, did not "just grow" but was shaped to respond to conditions of threat that were themselves far from the "objective facts" of war or surveillance that politicians and military officials made them out to be. The "bee with an electronic brain," in other words, was shaped by a set of specifically American assumptions about technology, "peacetime" war, and global control.

National Security Above and Below

The U.S. public did not innately know how to see aerial reconnaissance as national defense. Rather, they had to be taught, and this pedagogy of surveillance reached its zenith during the Cuban Missile Crisis. A *Time* magazine report from December 28, 1962 includes a brief remark by President John F. Kennedy: "The camera, I think, is going to be our best inspector."[47] The article, "Reconnaissance: Cameras Aloft: No Secrets Below," was published in the aftermath of the Cuban Missile Crisis, which brought the United States and the Soviet Union to the brink of nuclear war. On October 14, 1962, reconnaissance photographs of medium-range ballistic missile installations in San Cristobal, Cuba precipitated a standoff between the United States and the Soviet Union, resulting in an intense campaign to collect visual and electronic intelligence over Cuba.[48] Speaking about the discovery of the missile sites on national television for the first time on October 22, 1962, President Kennedy began by observing, "This Government, as promised, has maintained the closest surveillance of the Soviet military buildup on the island of Cuba. Within the past week, unmistakable evidence has established the fact that a series of offensive missile sites is now in preparation on that imprisoned island. The purpose of these bases can be none other than to provide a nuclear strike capability against the Western Hemisphere."[49] In Kennedy's opening remarks, the U.S. government's surveillance of Cuba becomes "unmistakable evidence." Earlier debates about strategic reconnaissance, including "Open Skies" and Francis Gary Powers's capture, positioned U.S. surveillance as potential aggression rather than neutral, objective "evidence." During the Cuban Missile Crisis, surveillance becomes unequivocal ("none other than to provide a nuclear strike capability") and Kennedy uses these "facts" to announce that "a strict quarantine on all offensive military equipment under shipment to Cuba is being initiated," as well as "the continued and increased close surveillance of Cuba and its military buildup."[50] The speech ties the "evidence" from surveillance to national protection, emphasizing surveillance as a defensive response to nuclear threat.

Kennedy's account papers over a number of secrets, however. First, the images he refers to were captured by early reconnaissance satellites, marking their newfound significance and potentially shaky accuracy. More importantly, Macrakis writes, "It was, in fact, a human source, the defector Oleg Penkovsky, the CIA's star GRU (military intelligence) source in the Soviet Union who first provided the CIA with this information. The satellite photo interpreters could decipher what they saw and provide additional visual evidence, but Penkovsky had given it context and meaning."[51] The significance of a human source is at odds with how the Cuban Missile Crisis and its aftermath were narrated to the U.S. public, wherein the camera seemingly provided transparent evidence free from the need of interpretation.

On October 28, 1962, the crisis ended when Radio Moscow announced that the Soviet Union would remove missile installations from Cuba in exchange for an agreement from the United States not to invade the island. The day before, an American U-2 pilot, Maj. Rudolph Anderson, was shot down by surface-to-air-missiles and died in the resulting crash, leading to another push for unmanned reconnaissance. Ryan Aeronautical had first presented the top-secret Firebee drone reconnaissance program to the air force two years earlier in 1960, and the program was funded in 1962. Based at the Atlantic Missile Range in Cape Canaveral, Florida during the Cuban Missile Crisis, Robert Schwanhausser, head project engineer, remembers how "the Ryan people were trying to figure out how their [reconnaissance] drone work might be continued. . . . Without a doubt they could do a job in Cuba."[52] However, from his position working in the Air Force Reconnaissance Division, Lloyd Ryan saw that "there was a great reluctance to deploy the system," and he recalled that General Curtis LeMay, the air force chief of staff, personally cancelled a mission to Cuba with the system in November. Ryan says, "It was due to the unknown nature of just how good it would be, and whether we were giving away a capability that we might want to save for bigger game."[53] After the Cuban Missile Crisis, Ryan Aeronautical was given the go-ahead to build thirty-eight new drones over the next two years.[54]

However, the secret mandate that led to the drone and the gamelike strategies emphasized by the engineers contrast with the public accounts of reconnaissance. *Time*'s "Reconnaissance: Cameras Aloft: No Secrets Below" explained America's strategy for national security and global control through the camera. Although the article suggests manned flights, "the camera" apparently acts on its own (perhaps, a reflection of the secret use of drones and satellites also used for surveillance). Referencing Kennedy's statement that the camera functions as the best inspector, the article explains, "The President's brief, blunt remark was deliberate understatement. For months the Cuban skies have belonged to U.S. photo planes—soaring, diving, circling, appearing and disappearing on swift, unexpected tangents. Diplomats may still argue about on-site inspection of Cuban missile bases, but the question is almost academic. Under the prying eyes of U.S. aerial cameras, Cuba lies as exposed as a nude in a swimming pool."[55] In this account, the flight appears unmanned, and cameras and planes act as if they have agency. The actions ascribed to the photo plane carry a curious collection of descriptors: soaring, diving, circling, appearing, disappearing, swift, and unexpected. They emphasize unpredictability and stealth, attributing a dizzying array of maneuvers to reconnaissance operations, which "expose" the ground below. These actions claim Cuban skies for the United States, bypassing territorial boundaries through the maneuvers of photo planes. The airspace above dominates the territory of Cuba below, made visible and open under the camera's "eyes." The "camera" claims to provide the United States access to the territory below. The article suggests that, unlike diplomatic arguments, the view from above makes discussion moot. The aerial camera exposes the land, in turn defining the territory below as known, inspected, and watched by the United States.

On February 6, 1963, the U.S. government offered an even more in-depth explanation of the new role of surveillance. John T. Hughes, a former geography teacher and an image analyst for the Defense Intelligence Agency, appeared on national television in response to doubts about the withdrawal of nuclear weapons from Cuba (figure 10). In an hour-and-twenty-minute presentation, he walks the viewer through a series of aerial images to show that no nuclear threat remained on the island.[56] The next day, the *New York Times* described the briefing as "one of the most unusual showings of such materials ever made by a government."[57] Previously, classified images from reconnaissance overflights had been shown to reporters or published in news reports. What marks the significance of this briefing, however, was that it showed television viewers—namely, the U.S. public—how to see the images from the perspective of a defense analyst. Through this pedagogy, the aerial image is defined as objective evidence and as the purported basis for establishing threat and attack.

Robert McNamara, Kennedy's secretary of defense, introduced the television briefing. He explains that the threat of a nuclear missile strike by the Soviet Union from bases in Cuba has been mitigated, stating, "It is our purpose to show you this afternoon the evidence on which we base our conclusions."[58] He calls on Hughes to show the aerial photographs that underpinned the military's conclusions. The briefing was probably not unlike the presentation Hughes had given earlier that day to the president and his advisers. The

FIG. 10 John T. Hughes conducts a military briefing for the news media showing the absence of Soviet missiles in Cuba with aerial reconnaissance imagery on February 6, 1963, in Washington, DC
Credit: Defense Intelligence Agency

New York Times report summarized the television segment by noting that the presentation "proceeded from prints showing open fields and woodlands last August and September, to the same sites in near-readiness for missile operations in October, and the same sites in dismantled condition today."[59] The photographs set the missile technologies against the island landscape, and their appearance and disappearance are tied to a natural view of territory—a point emphasized in the *New York Times* summary.

While Hughes is clear that U.S. intelligence missions have captured the images, pointing to the shadow of one of the government planes in a slide, he presents the photographs as if they speak for themselves, although the aerial views shown on home television screens would have been blurry and their features difficult to identify. The photographs use labels to indicate the critical features of the imagery, which Hughes points to in his briefing, teaching the viewer to see the photographs as if these explanations were an integral part of the landscape pictured, rather than an added layer of human interpretation. The view is both from the U.S. perspective and naturalized. The presentation claims to expose Cuba through the aerial photographs, know its potential to threaten the United States, and—through this very practice of seeing—defend against the enemy threat.

In Hughes's presentation, aerial intelligence is performed as if it were merely the viewpoint of the camera, providing evidence of what is happening on the ground below. The photo plane appears machine-like and autonomous, like the Firebee drone introduced at the outset of the chapter. The view from above aims to reassure U.S. television viewers with incontrovertible evidence that they are protected. The pictures leave out how the images are produced, contextualized, and analyzed (a process marked by indeterminacy, as suggested in the exchanges between the air force and Ryan Aeronautical about the use of the Firebee drone). Instead, the photographs suggest unmediated access to Cuba itself, as if the landscape passes directly through the camera to the viewer. Taken as an aerial picture, the political and military practices that produce the image of Cuba are made fact. Inscribed in this series of pictures are the geopolitical divisions between the United States and "enemy" territory. The camera view is framed as a nationalized view, which counters threat with evidence and an objective view of the world below.

4

Buffalo Hunter

● ●

The idea of the Cold War as a set of political, military, and industrial practices organized by automated control arose in tandem with postcolonialism. Nuclear missile buildup overlaid proxy conflicts between the United States and the Soviet Union extending to Asia, Africa, and Latin America, refashioning colonial occupation as military interventions, government assistance, economic aid, the academic fields that comprise area studies, and technoscience. The previous chapter emphasized how unmanning in the United States was used to promote automated, global control, while surveillance presented the U.S. public objective evidence of the enemy's activities and, thus, of a field of war. The discursive context around drone aircraft steadily shifted from naming the machine as "pilotless" or "manless" to "unmanned," displacing human action as if war were an automated sphere under machine, rather than human, control.

The previous chapter focused on drone research motivated by the Cold War. Unmanned aircraft, however, were never deployed to the USSR and rather fit within a context of proxy wars. The classified projects in the United States that used drones for reconnaissance explicitly drew on tropes of colonialism, iterating the destruction of Indigenous lands and histories as technological advance. These projects also drew on the illusion of automated machines, removing the colonizer from the act of continued violence implied by surveillance. Fantasies of empty landscapes for conquest, the erasure of Indigenous knowledge for universal technoscience, and the inevitable progress of civilization all figured into the rhetoric used by the secret reconnaissance projects built in the 1960s and 1970s. This chapter explores the intersection of these fantasies with the ontology of "unmanning."

The television and radio control systems built for the assault drones discussed in chapter 2 resulted in a surplus at the end of World War II. They had
been used to convert "war-weary" B-17 bombers to remote-control missiles. In
Operation Crossroads, a nuclear weapons test conducted in the Bikini Atoll
in the Marshall Islands between July 1 and July 25, 1946, both the U.S. Army
and U.S. Navy flew remotely operated aircraft that collected air samples and
photographic reconnaissance of the atomic explosions. They were among the
242 ships, 156 airplanes, 750 cameras, 25,000 radiation recorders, 5,000 pres-
sure gauges, and 4 TV transmitters deployed to the atoll, which involved the
participation of 42,000 personnel and spectators.

The stated purpose of Operation Crossroads was to study the effects of
nuclear attack on warships, equipment, and materiél, echoing interwar experiments with aerial bombardment at a heretofore unimagined scale. Operation
Crossroads aimed to simulate nuclear war. The use of drone aircraft for data
and image collection presaged their future use in the Cold War as surveillance
platforms. In the tests, these functions remained linked to earlier efforts to
use drones as prototypes of guided missiles. Operation Crossroads was also a
public spectacle and media event. Coverage of the tests continued for months
before and after the detonations, while reporters, members of Congress, representatives of the United Nations, and foreign delegations all attended.

Among the articles covering Operation Crossroads, an account of the drone
fleet was published in the *New York Times* on August 25, 1946: "The 'Drone':
Portent of a Push-Button War." The article explains, "Although one or two
planes were buffeted or damaged by the explosion, the crewless drones passed
through the atomic cloud, answered their 'mother's' commands perfectly, and
landed back at their bases." It concludes, "They are only the first category of
transoceanic missiles; rockets and robots, traveling at great altitudes and
speeds, will succeed them."[1] The article situates drone aircraft in the atomic
age, considering the system alongside "rockets and robots" as features of an
automated field of war. Despite the breathless description, the use of drone
aircraft in Operation Crossroads signaled an end to the projects developed
in World War II—though some pilotless flights were attempted for bombing missions in the Korean War. The article thus suggests how drone aircraft
formed part of a technological ideal that was in practice limited; remotely
piloted aircraft were still controlled by line-of-sight, meaning a "mother" aircraft with operators onboard was never far from the drone. The "push-button
war" was not automatic but orchestrated—not unlike the American pursuit of
global power—as if it were merely a set of technological outcomes.

In the Operation Crossroads test, the United States heralded technoscience as a means of global control. The mission juxtaposed the spectacle of a
nuclear experiment with the displacement of the island's Indigenous habitants.
Some of the remotely operated drones were part of the T-2 unit deployed by

the army, responsible for the collection of moving pictures and photographs of the atomic tests; they were assigned to the Air Photo Unit during Operation Crossroads. A large (25 inches by 18 inches), wooden-covered scrapbook from the unit, donated to the National Air Space Museum, is a souvenir of the mission. Inside, typewritten, mimeographed pages provide a brief history of the T-2 unit, followed by a collection of photographs from its deployment. The scrapbook records the final days of the people of Bikini Atoll on their land and its sublime destruction. The same photographic unit captured both sets of images. This photographic documentation purports military and technical advance, including drone aircraft, while repeated nuclear tests and radiation devastated the Bikini Atoll and underscored the postcolonial mission of the tests. Today, the more than four thousand descendants of the Bikini Atoll inhabitants are dispersed throughout the Marshall Islands and beyond.

The typewritten history makes note of the unit's commanders, preparations during the several months leading up to the test, and the massive amount of photographic equipment mobilized by the unit, remarking, "The number of cameras used in this operation parallels the immensity and importance of this mission. One-hundred fifty-seven motion picture and 106 still cameras ran through approximately 3,700 feet of film."[2] Much of the scrapbook comprises photographs of the equipment, including hundreds of cameras laid out on the airfield in Kwajalein. In the final pages, the drones are pictured showing "beeper" pilots operating the systems from ground-control units, followed by sequences of the "mother" and "babe" planes in the air (figure 11).

In the penultimate section, before a collection of photographs of the atomic explosion, there is a short part labeled "Natives." The images are a surprising shift from the previous photographs, which featured equipment and military personnel. Labeled "Bikini Marshall Islands May 1946," the first page of images shows the arrival of the U.S. military ships onshore, setting the fleet against the empty shores of the island, palm trees in the background. The next page, "Native Scenes," shows Bikini's inhabitants in the community's thatched structures. Women and children pose in the open doorways in one photograph, while an image on the bottom right corner features a large gathering of the community. The next page shows the U.S. military personnel conducting voice-recordings of the Bikini islanders. The following pages rely on tropes of ethnographic photography to document the island's people (figure 12), devoting two pages, for example, to "native children." The photographs provide few details—only one person is named, "King Juda of Bikini." The text in the scrapbook briefly mentions the pictures, noting, "Early in May a detachment of photographers was assigned to Eniwetok to cover the documentary history, maintain and install the equipment in the Drones and Mothers and to perform PRO [public relations officer] functions."[3] The U.S. military sought cooperation from the Indigenous inhabitants of Bikini Atoll

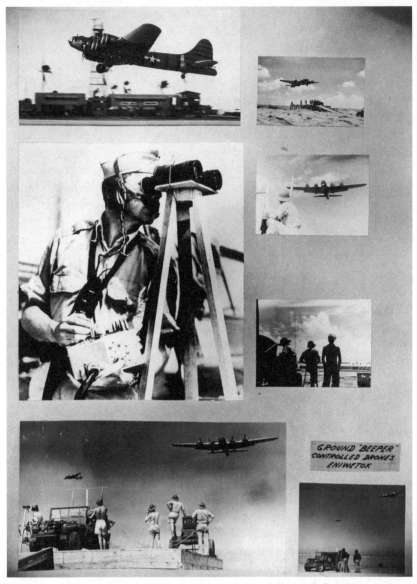

FIG. 11 Ground-beeper-controlled drones, Eniwetok, Marshall Islands, from *Operation Crossroads Scrapbook,* 1946
Credit: National Air and Space Museum

before the tests through assurances that their relocation from their lands would be temporary. Over seventy years later, 167 former inhabitants and their descendants continue to seek compensation from the United States for the loss of their land, means of subsistence, and way of life as a result of Operation Crossroads and the nuclear tests that followed over the next twelve

NATIVES BIKINI, MARSHALL ISLANDS MAY 1946

KING JUDA of BIKINI

FIG. 12 Portraits of inhabitants from Bikini, Marshall Islands, from *Operation Crossroads Scrapbook,* 1946
Credit: National Air and Space Museum

years. In total, twenty-three nuclear devices were detonated in the region between 1946 and 1958.

In the scrapbook, the short section of "native scenes" is followed by the iconic images from the nuclear tests, showing the massive, sublime mushroom cloud shot from an aerial perspective. It appears as a conclusion and a triumph. Yet, the first bomb missed its target, while the second bomb, which exploded ninety feet underwater, resulted in massive contamination. Irradiated water from the explosion fell back onto the vessels that were supposed to be destroyed, resulting in a fleet of contaminated ships in the ocean. Widespread exposure of personnel to radiation proved one of the most significant outcomes of Operation Crossroads. For the Indigenous inhabitants of the Bikini Atoll, the tests would prove devastating. For the tests, Indigenous inhabitants of Bikini Atoll were moved to Rongerik Atoll. The previously unoccupied island did not yield sufficient sustenance and they suffered from severe malnutrition. In the ensuing years, the former inhabitants of Bikini were relocated multiple times. In 1970, following assurances from the U.S. government that the area was free of radioactive contamination, around 160 persons returned to Bikini Atoll, during which time they experienced an elevenfold increase in radiation exposure. Inhabitants were evacuated in 1978. Their claims against the U.S. government for compensation are ongoing, while access to the Bikini Atoll is severely restricted and the island remains uninhabitable.

Images in the scrapbook enmesh drones with atomic weapons and an enormous array of instrumentation to collect data and imagery. The tests purported to study atomic destruction, though the effects of Operation Crossroads exceeded the parameters proposed for it and the third scheduled test never occurred due to concerns about radiation contamination. Operation Crossroads was widely recounted as a triumph of American technoscience and military power in 1946, although this evaluation has since been modulated due to the legacy of contamination. Pictures of the Indigenous inhabitants of Bikini Atoll in the scrapbook suggest how their erasure by the United States was paradoxically carried out by documentation. Including "Natives" as a section of the scrapbook shows how scenes from the island, and the collection of oral stories, prefigures not a record of Indigenous history but its erasure as part of American technoscience and global control. The images establish a narrative that aligns the removal of Native peoples from the island with the rise of the atomic age, culminating in scenes of a sublime mushroom cloud. Their inclusion is as a "historical document," as if the lives of Bikini Atoll's inhabitants were part of the past, written over by technoscience, a brief moment in between the rise of drone aircraft and nuclear weapons. The possibility of their return to the island appears foreclosed before it was declared permanent and the extent of radioactive contamination known. Yet, the photographs also insist on an ongoing struggle in the present against a total domination imagined by nuclear destruction, which continues through the redress sought by the Marshallese against the U.S. government.

During the Cold War, the United States disavowed colonial territorial expansion, claiming to support anti-imperialism and decolonialization. Such pronouncements were repeatedly undone by examples like Operation Crossroads. In "Islands: The United States as Networked Empire," Ruth Oldenziel outlines how Cold War technopolitics relied on the acquisition of islands to control vast areas of the globe. The strategic use of islands was premised on pre–World War II naval power. Oldenziel writes that in the Cold War, the "legal and geopolitical mobilization of islands into the American orbit took a specific technological shape that would dominate the postwar era: emptying out space by filling the islands with technologically intense systems that obscured the political imprints of the United States."[4] The use of islands to build U.S. global power was predicated on a disavowal of their political uses.

Unmanning develops in this context, as the U.S. government flew reconnaissance missions in the 1960s and 1970s while outwardly denying their political aims. Instead, these classified missions were intended as a secret intervention in the global sphere. A dispersed network of bases, laboratories, industry, and personnel allowed for the illusion of unmanning to cohere, shaping a context for U.S. global control that claimed to be machinelike, deterritorialized, and all-seeing. Yet, this global control is only made conceptually possible by erasing the land below. In this chapter, I address how those erasures extend from the American West to bases throughout the Pacific and are further negated by the U.S. government via their classified status. The programs, moreover, use the vocabulary of colonialism to outline their exceptional status, funding drone reconnaissance through "Big Safari" and naming a mission in Vietnam "Buffalo Hunter."

Achille Mbembe writes, "The colonies are the location par excellence where the controls and guarantees of judicial order can be suspended—the zone where the violence of the state of exception is deemed to operate in the service of 'civilization.'" The automation of war, tested by Operation Crossroads, is interwoven with "colonially generated fantasies of wilderness and death and fictions to create the effect of the real." These erase local claims to land and Indigenous histories, replaced by "civilizing" practices of technoscience that rely on the suspension of order. The result is a proliferation of violence. As vertical sovereignty increasingly defines the terrains of this violence, Mbembe argues, "Everywhere, the symbolics of the top (who is on top) is reiterated. . . . Various other technologies are mobilized to this effect: sensors aboard unmanned air vehicles (UAVs), aerial reconnaissance jets, early warning Hawkeye planes, assault helicopters, an Earth-observation satellite, techniques of 'hologrammatization.' Killing becomes precisely targeted."[5] Unmanned reconnaissance is made by and makes the symbolics of the top. The term overlays aerial photography with colonial violence and states of exception. Unmanning claims to capture a transparent, top-down view of the world below. Yet the pictures captured by its cameras are organized by networked erasures that reiterate colonial practices and the exceptional use of state power. They register the rise of technoscience and imagine the world below as a landscape to be captured. These experimental trials also connect the failures of unmanned aircraft and the impossibility of the total view that is imagined for it. Rather than truly existing as all-encompassing, unmanning is made by networked parts that shape the field of war the drone claims to document and, inadvertently, point to the drone's limits and the power imagined by its symbolics. Technical advances in late modern war exceed the objective and automatic frameworks it claims, drawing on failed projects to reiterate—and fragment—the view of the ground below. The conditions that make the symbolics of vertical sovereignty are enmeshed with prosaic military secrets, failed experiments, and the continuous erosion of "being on top."

"Big Safari": The Camera and the Gun in the American West

As a target drone used for air defense training, the Firebee located unmanned flight in the "desolate" American West (also discussed in chapter 3). Images from a 1953 press release show the Firebee flying over New Mexico, with the iconic White Sands Desert and Organ Mountains in the background (figure 13).[6] The caption, written by an industry publicist and pasted on the back side of the photograph, reads, "Speeding on its lonely way over the desolate sands of New Mexico, the Ryan 'Firebee' pilotless drone presents a spectacle as eerie as an uninhabited missile from another planet. Linked with human intelligence by electronic radiation, the obedient Q-2 responds to commands from a remote-control ground station until its fuel is exhausted and an ingenious parachute recovery system brings it back to earth."[7] The bulbous head of the pilotless plane, the squat shape of its fuselage, and the short wingspan all distinguish the aircraft from the drones that came before. Unlike drone targets tested by human "safety pilots" in World War II or the converted B-17s used in Operation Crossroads, the Firebee was designed without a cockpit, and no human ever flew onboard.[8] The drone was short, only twenty-two feet long. The caption portrays the drone "as an uninhabited missile from another planet," attributing alien qualities to the unmanned aircraft and its "strangeness." Identified with the military designation Q-2, it is also "obedient," responding to transmitted commands, though the drone's "lonely" flight over the "desolate" desert suggests it flies on its own.[9] Operators remain a spectral presence, linked to the drone by "electronic radiation" and "intelligent" signals, setting out a haunted disavowal of the human. The image and caption set out the contradictions of unmanning: alien and human-controlled, lonely and obedient.

In the photographs of the Firebee from 1953, the sand dunes in the empty desert, an "uninhabited planet," overwrite the military and political formations in the land below. New Mexico became a U.S. state through the seizure of lands both from Native Americans and Mexico. White Sands was territory of the Mescalero Apache through the end of the nineteenth century. The Mescalero resisted Spanish and American colonial settlements until 1873 and continue to play a prominent role in the region, living on a reservation located fifty miles

FIG. 13 Ryan Firebee pilotless drone over Holloman Air Development Center, NM, circa 1952–1953
Credit: National Air and Space Museum

from the military base. The drone fills an "empty" landscape that was never uninhabited but, rather, has been characterized as such by the United States. In 1941, public land grazing licenses around Alamogordo, New Mexico (a planned community built to support the El Paso and Northeastern Railroad) were cancelled and a large-scale military base was built. This repurposing of public lands, previously seized from Native peoples, occurred throughout the American West during World War II.[10] Designated as the Alamogordo Bombing and Gunnery Range in 1941, the area would be renamed the White Sands Proving Ground in 1945. The name change came just prior to the Trinity test held at the site on July 16, 1945, when scientists from the Manhattan Project detonated the first atomic bomb in White Sands Desert.[11] The region served as a missile test site throughout the Cold War and continues to be closed periodically for this purpose; the systems are still tested against Firebee targets. The base was also a test site for the drone's secret conversion to unmanned reconnaissance. Today, the nearby Holloman Air Force Base is the primary training center for pilots and sensor operators of MQ-9 Reaper unmanned combat air vehicles.

The air force funded the conversion of a Firebee target drone for reconnaissance collection in 1962 through a program called "Big Safari," set up ten years earlier. In Donna Haraway's essay "Teddy Bear Patriarchy," she notes how the camera and the gun become interchangeable in the hunt for big game, while the safari claims a "natural" hierarchy that emphasizes the preeminence of white men and the erasure of Native knowledge and labor.[12] Although I could find no documented rationale for the program's name, its mission statement notes that Big Safari funds aircraft "of sufficient importance and priority to warrant preferential treatment, quick reaction and extraordinary procurement procedures, in other words, 'big game.'"[13] While the program's mission statement emphasized the exceptional status of the air systems, the name may have been largely aspirational, as reconnaissance aircraft were typically a lower priority for the air force than jet fighters or missile systems. Funded outside standard procurement, experimental reconnaissance was developed in a patchwork manner. Big Safari can be taken as a metaphor for the project's outsized ambitions: to secure the place of aerial reconnaissance as all-seeing control, with Haraway's camera replacing the fighter jet's gun.

The Lightning Bug drones developed by Ryan Aeronautical from its Firebee target plane were the first *unmanned* reconnaissance systems financed by Big Safari. "Lightning Bug" was only one of many codenames, which were changed multiple times during the next thirteen years in an effort to maintain the secrecy of the project. The contract given to Ryan Aeronautical on February 2, 1962, called for the conversion of four drones into photoreconnaissance systems in ninety days, leading to the experimental trials discussed later in this chapter. Today, the program Big Safari continues under the same name and is used for the rapid conversion of aircraft. Notably, in 2000, the program was used to weaponize the reconnaissance RQ-1 Predator drone, completing a prototype that was deployed in Afghanistan sixteen months later. The conversion of the RQ-1 Predator was the first weapons technology funded by the procurement program. These developments instantiate the claim long held by advocates of unmanning: that a weaponized drone was a "natural" outcome the projects. The stories I discuss work against this logic, however, showing how the drone's development and use emerge in specific circumstances that were not preordained.

Working within the Big Safari time constraint, Ryan Aeronautical engineers relied on the design of the Firebee target to guide the development of the reconnaissance drone, extending the length of the target drone to just over twenty-five feet. The most significant modification was to the navigation system. Previously, the Firebee flew based on a preset course of the altitude, distance, and direction to be traveled. The first new navigation unit for the reconnaissance system was built using a timer-programmer from a telephone stepping-switch and a gyrocompass, which guided the direction of the aircraft over set time intervals. This system would later incorporate Doppler radar. Other modifications to the target drone aircraft included a thirty-five-inch section spliced into the fuselage to carry sixty-eight additional gallons of fuel, extending the range of the aircraft to over one thousand miles. The nose of the aircraft was modified for a Hycon camera, also used in U-2 reconnaissance planes. Two large film magazines would unspool the film in opposite directions, which maintained equilibrium in the aircraft. The camera system captured pictures on large-format negatives as the film unrolled during flight.

Robert Schwanhausser, lead project engineer for drone reconnaissance at Ryan Aeronautical, remembered the 1962 project in an interview in 1971. His comments suggest the day-to-day ways that the engineers treated the test flights between Holloman Air Force Base and Wendover, Utah as a "safari" for pictures. One story involves the decision to use the Hycon camera (as opposed to a mock-up) in the first experimental, long-distance flight on April 20, 1962. As with previous drones, the system was flown from a ground-control station and a launch plane. Schwanhausser recalls, "The Lt. Colonel who was in charge of the drones at the time felt very responsible about this high-priced so-called piece of equipment."[14] Schwanhausser and the government in-plant representative for the project, Jim Regis, discussed their options. "We decided that with all the chips being down, let's go for broke. If we lost the bird,[15] hell, we lost the bird, but if we lost the bird and still had some good film, we would prove our point and we just wanted to prove we could take some pictures."[16] His statements set out an exceptional quality to the secret project, describing the flight both as a gamble and as proof of the reconnaissance drone's worth inscribed in "some good film."

Just prior to the midair release of the drone from the DC-130 launch aircraft, the lieutenant colonel—who had ordered the Ryan Aeronautical engineers not to use the real camera in this first test—showed up. Schwanhausser recalled that at the time they "were having a wonderful flight, it was a marvelous flight, everything was just great." Meanwhile, the colonel left and went straight to the base commander. "We were notified that we were going to be thrown off base, that it had better be a damn good flight, because it was going to be our last flight." Traveling 761 miles to Wendover, Utah, the drone completed the round-trip flight in ninety-eight minutes, monitored continuously from a launch plane that followed the drone in the air and by ground tracking. Schwanhausser remembered, "We did continue to fly at Holloman. [And in] the meantime we created a tremendous scene."[17] The difference was eventually patched over by commanders in Washington, DC.

On May 3, 1962, the same reconnaissance drone was flown for its third trial flight. On the return leg of the flight from Wendover, Utah, a momentary loss of connection between the drone and the altimeter tracking system caused the automatic parachute recovery system on the drone to be activated. The drone landed between two peaks in the Magdalena Mountains, 120 miles from White Sands. The drone crash potentially compromised the secrecy of the reconnaissance drone. A military helicopter was sent to pick up the drone, but with the wind and high altitude, the helicopter did not have enough lift, and it also crashed. Schwanhausser recalls, "Now we've got both an Army helicopter and a classified drone with all its secret gear splashed on the mountain side."[18] The air force recommended the drone be burned in the interests of security. At the time, however, there were only two working models of the unmanned system. Ryan Aeronautical instead sent a crew of its civilian employees "armed with guns in their holsters as a security precaution" to retrieve the drone. "By four o'clock we had completely disassembled the bird. Two Air Force trucks had broken down trying to get to the site. In the meantime we had 'conned' an Army sergeant with a weapons carrier into helping haul some assemblies out. By later afternoon we had the bird sitting in a meadow, some of the assemblies resting on old tires so they wouldn't get banged up."[19] The landing and disassembly of the drone occurred over the weekend. By Tuesday, the system was ready to be flown again. Schwanhausser recounts the next flight: "We flew it over the same area and took stereoscopic pictures of the chopper that had crashed into the mountains."[20]

Schwanhausser's description of the Lightning Bug's test flights situates the drone neither in a context of automatic control nor in the objectivity associated with the drone's-eye view but, instead, conveys investment in the project. The engineer's characterization portrays the flights as a gamble and a hunt for images; the crash is an opportunity to insist on the functionality of the system, repeating its flight as a success. These exceptional qualities are tied to the secrecy that surrounds the project. Speculation about unmanned aircraft being used for purposes other than target training circulated before the beginning of the project. In January 1968, White Sands Missile Range inadvertently mentioned the classified drone reconnaissance tests in an end-of-the-year press release of base activities, giving away the codename for the project, "Firefly"—another name for a secret unmanned reconnaissance system. The report explains, "Firefly will test a special purpose vehicle. Initial flights will originate over the White Sands Missile Range with later flights originating from above the Pacific Ocean and terminating over White Sands Missile Range."[21] This was the first public mention of the reconnaissance drone.

The next year, on August 6, 1969, a headline in the *Albuquerque Journal* read, "Secret Something Falls to the Earth." The article remarks on "the emergency descent by parachute of a super secret unmanned aircraft in full view of Los Alamos residents" and describes the drone "dangling from a bright orange and white parachute" as it eventually landed "behind the fences of a security-conscious Los Alamos Scientific Laboratory technical area." The article notes that "reverberations from the thud of the graceful bird's landing on the northern New Mexico plateau, 150 miles away from White Sands, were felt all the way to Washington."[22] Attempts to keep the reconnaissance drone secret were splashed into headlines, as an Associated Press wire article ("Sudden Landing

Unveils New Drone") put together the inadvertent press release from White Sands about Firefly with the drone landing at Los Alamos.

Ryan Aeronautical engineers and military personnel addressed the incident and the behind-the-scenes actions to manage the crash at Los Alamos and explain its exceptional characteristics. At the Holloman Air Force Base ground-control unit, panel lights signaling a control failure came on after the drone had been flying for three hours. The only option for the controllers on the ground was to "hit the 'panic button,' which released the parachute . . . permitting it to descend with minimum damage so that it might be recovered."[23] Art Rutherford, who was aboard the DC-130 launch plane that tracked the flight, later recalled that "everyone was looking out the windows to see if they could spot the bird. We were flying over an area of deep arroyos [gullies], with a big complex of factory building or hangers in between." Rutherford remembers arriving at the Atomic Energy Commission in Los Alamos: "Our Special Purpose Aircraft had landed right in the middle of the driveway, the right wing skidding under the guard rail. . . . The news people were there, pressed up against the perimeter fence or up in trees to get a better overview."[24] Employees of the Atomic Energy Commission at Los Alamos, familiar with classified operations, asked few questions about the project. The incident, however, might have caused a much greater disaster: "The drone was dripping fuel as it came down, drifting over the main plutonium processing plant. Had it gone down there they would have had to evacuate the area."[25] After the incident, the air force acknowledged the existence of the Firefly for reconnaissance, though it provided few details.

The account of the unmanned reconnaissance system coming down over Los Alamos links two threads of Cold War technoscientific research in New Mexico: unmanned aircraft and nuclear power. The confluence juxtaposes two of many efforts to produce secrecy in the Cold War United States, pointing out how the technical advances that were claimed for many military programs were enmeshed with classified silences justified as in the best interests of national defense. The secret qualities of national security defined the United States against an external threat, while the projects produced internal risk and division in the process of maintaining their secrecy. Writing about classified weapons development in Los Alamos, Joseph Masco notes, "In striving to keep the internal . . . costs of the national security state invisible, state secrecy both produces and enforces an official fiction; namely, that the only legitimate forms of nuclear security and risk are located outside the territorial borders of the nation state."[26] He ties the risks to the disproportionate effects of nuclear contamination experienced by racial minorities who labored in the nontechnical positions at the laboratory. The drone crash at Los Alamos did not create a nuclear disaster. Yet, the close call highlights how protections proposed by the nation state use secrecy to erase the exceptions suggested through Big Safari; its failures are described as a measure of technical advance rather than a risk. The veneer of secrecy disavows state investments, at once political and economic, in the project as instead technological experiments in an empty landscape.

"Honey" from the Bees: Drone Reconnaissance in Vietnam

Chapter 3 outlined how the U.S. public was taught to see reconnaissance as national security, through claims that the aerial view relayed facts on the ground for national protection. Yet, secret missions also organize aerial reconnaissance in equal measure, cementing the "symbolics of the top" in the discourse of unmanned flight and its top-secret use. Caren Kaplan observes that histories of remote sensing typically describe "the revelation of a stable view, made incrementally and more precisely legible through progressive technical innovation." She goes on to assert that these accounts "can be read against the grain to remind us that a visual culture is always in the process of being pulled together even as it never quite holds true."[27] Networked practices of secret aerial reconnaissance "pull together" a visual culture that establishes a power dynamic between above and below, stylized as an objective image. Aerial reconnaissance—for both manned and unmanned flight—relied on the organization of flights, photographic laboratories, and the labor of technicians and analysts.

This section focuses only on the capture of aerial photography by drone flight to emphasize the contradictions between what the flights aimed to produce—a transparent, objective view of the ground below—and the network of humans, machines, and media mobilized through unmanned flight that produced this context, all the while never seeing the images they captured. The account explores what is at stake in the shift from manned to unmanned reconnaissance, outlining numerous challenges faced by the Ryan Aeronautical engineers. Like the drones' flights over New Mexico, which are described with colonialist rhetoric that stresses hunting in an "uninhabited" landscape, the flights in Southeast Asia dramatize the symbolic binary between above and below in the context of U.S. postcolonial power and counterinsurgent responses in Vietnam.

In response to American naval engagements with North Vietnamese torpedo ships in the Gulf of Tonkin on August 2, 1964, Congress authorized President Lyndon B. Johnson to use U.S. military forces in Southeast Asia.[28] Immediately following this authorization, reconnaissance drones were deployed to the region. The reconnaissance drone system was rebuilt according to at least two dozen configurations, with contracts ranging from a few aircraft to hundreds between 1962 and 1975. Thousands of images were captured by drone aircraft, although to this day only a few have been released to the public. Promoting the use of unmanned aircraft in Vietnam, Ryan

Aeronautical claimed the converted drones collected "quite a bit of 'honey' in the way of intelligence-gathering during reconnaissance missions in Vietnam."[29] Their "honey" both calls back to the original image of the drone as a faithful "bee," from chapter 1, and the description of aerial reconnaissance that John Hughes provided for the U.S. public: images as objective evidence safeguarding the nation, discussed in chapter 3. Personnel who flew the missions did not themselves have access to the images they collected, which were instead analyzed by military intelligence. A Ryan Aeronautical employee, Ed Sly, explains that "while we didn't see the results . . . we understand the camera brought back significant information."[30]

The shift to unmanning met resistance. Many in the air force considered drone reconnaissance cumbersome and error-prone, while its strongest advocate was the military industry that built the systems. Engineers from Ryan Aeronautical received word on August 4, 1964 that the high-altitude reconnaissance drones built by the company would be deployed as part of the military buildup for the Vietnam War. As an industry contractor, Ryan Aeronautical was central to the operation of unmanned reconnaissance in the Pacific. After all, it did not merely design and sell the systems but also trained the air force in their use. Ryan's role as a contractor reflects shifts to military industry in the Cold War and the newfound importance of industry and academic contracts to weapons innovation.

Alex Roland observes that within military industry in the Cold War, "The test of a weapons system was not its parity with the weapons systems of enemies or potential enemies but rather parity with the next generation of weapons systems that industry could envision."[31] Schwanhausser remarks that the biggest challenge in the early days of the "unmanned" drone reconnaissance program was "a human relations problem." He says, "We wanted the bird to look good and we wanted the Air Force to look good and we needed to have a lot of cooperation." As lead project engineer, his interest in the program's success was perhaps not always shared by air force personnel assigned to the projects. Schwanhausser notes that during the first eighteen months of the program, Ryan employee Dale Weaver "flew practically every mission and he ended up flying more combat missions than any Air Force guy because they were rotating and we were just leaving Dale there."[32] The work was monotonous, and the missions relied on a network distinct from that of base, aircraft, and pilot typical of the manned flights.

For instance, the navigational system of the drone operated through a series of plotted points, preset prior to each reconnaissance mission. Below are two accounts: the first from an official U.S. Air Force history of drone aircraft and another from a letter by Schwanhausser. The first outlines the idealized, mechanical function of the aircraft, while the second emphasizes the work necessary for the programmed flight. The air force history explains, "Drones had self-contained guidance systems consisting of a programmer compass, Doppler equipment, and an autopilot. Before each mission, operators programmed each drone's system to guide the drone from its launch point along a pre-planned track over the reconnaissance targets, then to a recovery area."[33] The description accounts for the drone's flight based on a series of planned points. According to this report, the Doppler system checks the location at seven-mile intervals through the radar signals, while an altimeter maintains the aircraft's altitude. In case the system loses connection to radar, the drone is also programmed by dead reckoning, which would direct the flight based on the distance traveled over time. In a letter from October 2, 1964, Robert Schwanhausser explains preflight operations: "By approximately 3:30 a.m. the operations crew arrives from the Command Post and commences programming the birds. This is done with USAF people on one bird and Ryan on the other. One man reads the program while another patches it. Then the two fellows change places and the patcher reads while the other checks. Then the Ryan and USAF crews exchange birds and check each other. In addition each crew has done their own planning and program calculations independently and then cross-checked each other."[34] Programming was labor intensive and would last several hours, layering together multiple inputs in the process of establishing the flight path for the reconnaissance missions. The drones were usually launched between seven or eight in the morning, after three to four hours of setting the navigation program. Patchers checked, rechecked, and crosschecked their work, inadvertently highlighting the potential for human error in the process.

Programming was not the only challenge to navigation. The series of preplanned points that were to guide each mission often resulted in some degree of loss of direction during flight, which could be due to weather conditions and other external factors as well as equipment failure. For example, during a flight on September 21, 1964, "the course was about 2 degrees to the left of the intended course. . . . After the first leg it rolled out onto its new bearing about 8 n.m. short but on a perfect heading." Or, four days later, "the bird was dropped in a very poor launch area due to navigation difficulties with the launch aircraft." And on September 29, 1964: "The first 60% of the leg appears to be 2 degrees left and then changes to an almost convergent heading. The drone was 34 n.m. long on this leg and our information shows a very slow, wide turn."[35] In total, this model flew seventy-eight missions and averaged only 2.6 missions per vehicle before the aircraft was lost or broken beyond repair.[36]

Drone reconnaissance was initially based in Okinawa, Japan, at the Kadena Air Base in 1964. Upon receiving orders from Air Force Strategic Air Command (SAC) in Omaha, Nebraska for a flight, technicians prepared the unmanned aircraft at Kadena and launched them from a Lockheed DC-130

Hercules plane fitted with pylons, which would release two drones over Southeast Asia and China during its flight. The unmanned aircraft were then monitored by personnel in a "radar van," based in Taiwan, which served as a mobile ground station to track and recover the reconnaissance aircraft. Drones were picked up by helicopter after they descended by parachute in Taiwan, and the film would be shipped via courier jet to be processed back at SAC in the United States. The reconnaissance drone would be shipped back to the base in Japan for another flight, if the aircraft was not lost or broken.

The first mission flown from Kadena Air Base to Taiwan went poorly. It also indicates how the networks of parts that made unmanned flight possible were envisioned as a closed system distinct from an outside world made strange. The following account demonstrates how, when the machine failed (as it often did), the human part of the drone network deemed those not included in the network as "strangers" appearing out of nowhere, just as the "desolate" land in New Mexico was thought to have no one in it. "The bird landed in a rice paddy so the impact switch wasn't triggered, the chute did not disconnect and the bird was dragged until it flipped on its back. An Army helicopter came in and picked the bird up in a hurry as we were attracting a lot of strangers who appeared out of the bushes and woodwork."[37] The offhand remark about "strangers who appeared out of the bushes and woodwork" contradicts the alliances between the United States, Japan, and Taiwan that make unmanned flight possible. Instead, the way the drone opposed "above"' and "below" in part structures the conditions for its handlers to conceive of its operation as taking place over "a lot of strangers." The persons in Taiwan were not strangers at all, of course; they are the locals who lived and worked in the rice paddies where the drone crashed. These divides between foreign and domestic, stranger and Native become enmeshed with the networked parts of the drone, presupposing a group of secret insiders with everyone else made strange.

In the Cold War United States, military and political developments operated on the assumption that technology makes "the world so small that it can be overseen and managed easily."[38] Strategic reconnaissance was critical to this formulation and constituted a military strategy that operated through seeing and watching as if from an all-seeing, objective "above." Drone reconnaissance in the Cold War, however, shows how the significance attributed to "high ground" is produced and enacted. The drone programs moved forward and acted as if they could dominate from above not because the technology does, in fact, dominate—the reconnaissance drone project is widely deemed a failure—but because the system is animated to act *as if* this were the case. By insisting on the reality of a fragmented, incomplete, and partial aerial view, "being on top" is constructed in military-industrial propaganda and bolstered by a structure of secrecy to hide its own limits. The divide between above and below implied by the framing of aerial pictures takes on meaning, even when no images are seen by the drone mechanics, pilots, or the U.S. public.

Exporting Colonialism: The Buffalo Hunter Report

While manned reconnaissance captured much of the aerial intelligence over Vietnam, increased anti-aircraft defenses and the challenges of monsoon weather continued to bolster the case for unmanned drones. One month into its mission at Kadena Air Force Base in Okinawa, Japan, the drone reconnaissance unit was transferred to Bien Hoa, South Vietnam. During the next eighteen months, the system was redesigned, leading to the production of twenty-eight more drones. In 1969, the instrumentation company Teledyne acquired Ryan Aeronautical, which became Ryan-Teledyne. One of the largest contracts received by Ryan-Teledyne was for a low-flying reconnaissance drone, flown between 1970 and 1972 in top-secret missions known as "Buffalo Hunter." In over one thousand missions, the drones attempted to track more than thirteen thousand high-priority targets and captured photographs of close to five thousand sites, with an approximately a 40 percent success rate.[39] The aircraft flew close to the ground, beneath the cloud cover that obscured the battlefields in Vietnam from other forms of aerial reconnaissance, though the majority of the flights flew off course. By the end of the Vietnam War, all of the air force's drone reconnaissance programs were cancelled and the experiment with unmanning was widely deemed a failure.

On July 24, 1973, a confidential report about Buffalo Hunter was submitted by air force Maj. Paul W. Elder as part of Project CHECO (Contemporary Historical Examination of Current Operations). The opening pages explain: "The counterinsurgency and unconventional warfare environment of Southeast Asia has resulted in USAF [U.S. Air Force] airpower being employed to meet a multitude of requirements. These varied applications have involved the full spectrum of USAF aerospace vehicles, support equipment and manpower."[40] Positioning unmanned aircraft as counterinsurgency technologies, the report differs from earlier portrayals that had emphasized the protections unmanned aircraft provided for pilots and reduction of political risks involved in reconnaissance. Now, rather, unmanned spy planes counter the "unconventional" warfare environment in Vietnam.

The tone of Elder's report, which evaluates the systems for future uses, also differs from the personal and technical accounts of the projects made by Ryan Aeronautical engineers. It aims to articulate the system's effectiveness in gathering intelligence and how this information was used in the "warfare environment." However, there is a moment when the framework of objectivity cracks and Elder is almost joking with the reader, as he addresses the name of the project:

These reconnaissance operations functioned under tight security; and to maintain that security the reconnaissance directors changed the nickname of the operation several times. . . . By the BUFFALO HUNTER era, however, the drone's use was no longer a tightly-held secret. Howard Silber in an Omaha *World-Herald* editorial said that the "Buffalo Hunter can spot a water buffalo standing belly-deep in the muck of a rice-paddy." Although water buffalos were hardly the reconnaissance target for the drones, Silber's wry assessment of the capability is an accurate one.[41]

This passage highlights the intersection of colonialism and secrecy brought together by unmanned reconnaissance to produce the symbolics of the top. The name for the operation, Buffalo Hunter, ties homegrown U.S. settler-colonialism to its overseas incarnation via the drone's capacity to target water buffalo in Vietnam. Elder elides the actual targets of the reconnaissance, the North Vietnamese, using the animal hunt to substitute for the military and political conflict that itself is arguably colonialist in nature. The passage also attested to the accuracy of drone reconnaissance, although, overall, the report outlines the limitations of image collection and the failure of most unmanned missions to capture photographs of the targets on the ground that they were deployed to monitor. By first associating the system with a buffalo hunt and then implying that the reader of the report would share a similarly colonialist point of view and thus be in on the joke, the report fits the drone into a framework of U.S. colonialism.

Yet, the use of drone aircraft in Vietnam was far from a totalizing expansion. Producing the pictures was complex, and the missions could not provide time-sensitive information. Reconnaissance drones were first based in Bien Hoa Air Base in South Vietnam and later in Da Nang Air Base in Thailand. Film was flown to a separate laboratory for processing. Track planning and targeting for the flights was done by SAC in Omaha, Nebraska. The "Mission Planning" section of Elder's report details the consequences of these divisions, explaining that time-sensitive targets unfortunately had to be approved by SAC in Omaha. Air divisions active in the region wanted direct control of the project and found the organizational bureaucracy to be a problem. For example, "[General Lavell] described the target request procedures as 'cumbersome, time-consuming, and insufficiently responsive to urgent requirements to develop or revise BUFFALO HUNTER missions in response to changing threat and weather conditions.' "[42] Aerial reconnaissance did not respond to the changing temporalities of war and the environment.

The reconnaissance drones flown for Buffalo Hunter missions used a camera that captured 180 degrees of lateral coverage on the ground, exposed on 1,800 feet of 70-millimeter film. At 1,500 feet in altitude, the camera was capable of capturing 120 nautical miles of continuous photographs. However, the enormous, ostensibly continuous photographic strips produced by the drone produced partial and fragmented views of the ground below. The company history of the Ryan project includes two strips of photographs from the 180-degree lateral shots. They are among the few declassified reconnaissance images captured by drone aircraft in the Vietnam War.

The first photographic strip captures the roofs of village houses, small ponds of water, and agricultural fields in the distance. It is cropped to expose the anti-aircraft installations in the top quadrant of the photograph. This first reframing shows the military site adjacent to the fields, while the final enlargement shows only the batteries (figure 14). Like the photographs of the Cuban missile installations, it is numbered to highlight seven operational guns at the site. The final version of the photo asks the viewer to see only threat.[43] The image was also used to show how North Vietnamese troops were embedded in civilian contexts. The fragmented image cuts out what the military deems most important: gunnery that would be used to attack U.S. aircraft. The original photograph strip might be read differently, however, connecting the military installations, agricultural fields, and houses—a juxtaposition obscured in cropping the image to show only anti-aircraft gunnery. The broader picture, rather than showing threat alone, highlights how aerial attacks against Viet Cong killed civilians in their villages and fields.

Elder observes, "The North Vietnamese were undoubtedly familiar with the dwarf aircraft that regularly buzzed their cities, airfields, rail lines, bridges, roads and waterways."[44] Set in the context of everyday life in Vietnam, the drone is minimized, described as a "dwarf" and characterized by its buzzing hum rather than by the hunt metaphor in the project's name. For the U.S. public, as a component of national security, the production of aerial images must remain a secret. Yet, the aircraft in Vietnam are visible elements of wartime infrastructure to the "strangers" who witness them and the networks of humans that operate them. The second photographic strip included in Ryan's company history looks back into the erasures of aerial targeting. "On one mission in October 1968 when the low-altitude control system wasn't working properly, [the reconnaissance drone] came back with a truly spectacular picture. Instead of flying about 1500 feet over North Vietnam it came in at about 150 over the terrain, flying under a major power transmission line and taking a remarkable series of photos in which natives in coolie hats can be seen on the road staring up at the drone overhead."[45] The image is a panoramic shot captured by the drone as it flies under power lines with an electrical tower at the top of the photographic strip. A number of people walk on the road below (figure 15). It is unclear if they are really "staring up." From above, Ryan describes the people using colonial language, "natives in coolie hats." The image indicates how aerial views, particularly the 60 percent of drone flights

FIG. 14 Horizon-to-horizon photograph from a low-altitude reconnaissance drone; the enlargements show one-third of the original photograph and a detail of the anti-aircraft batteries, October 6, 1968
Credit: U.S. Air Force

FIG. 15 Horizon-to-horizon photograph from a low-altitude reconnaissance drone flying beneath high-tension electric lines, October 6, 1968
Credit: U.S. Air Force

that did not find their assigned target, are partial and particular—one might even say, misguided—in their gaze.

The people in the image who may or may not be looking back at the drone trouble the claim that surveilling the world proves that it can thereby be easily managed. When declassified and juxtaposed with the seemingly objective accuracy of images of Cuba promoted to the U.S. public, these pictures demonstrate different ways of knowing and seeing. The practice of aerial surveillance via drone produced changing territorial configurations, tied to how reconnaissance imagery simultaneously protects, monitors, and oversees the land below. In spite of this power to reconfigure how territory is seen and thus understood, the war in Vietnam showed that seeing does not guarantee control or knowledge. Aerial reconnaissance is presented to the military and the U.S. public as a way of protecting the United States from threat and of registering an objective view of evidence—supposedly produced automatically—without the human hand. This formulation is layered over the actual practices that make reconnaissance flight possible, which interconnected a wide web of military and industry personnel who would not necessarily have ever seen the images that were the result of their missions. Producing the image, moreover, often required multiple cross-country flights or networked operations between various countries. Day-to-day details of production reveal the ways in which the practice of global surveillance and the images it makes continue to be fragmented and partial. These practices situate an aerial view of the ground below within a milieu of U.S. colonialism, emphasizing an image that erases the day-to-day lives of the people who live there.

5

Pioneer

● ●

In 1986, a secret report by the CIA entitled "Remotely Piloted Vehicles in the Third World: A New Military Capability" considered the emerging roles of unmanned aircraft. Rather than assert remotely piloted vehicles (RPVs) primarily in the Cold War arms race between the United States and the Soviet Union, the report proposed their import for conflicts in the "Third World." In other words, the report situated the aircraft in a postcolonial context rather than aligning it with the global superpowers that had dominated the Cold War. The primary basis for its claim was the "successful use of [RPVs] by the Israeli military during the 1982 Lebanon conflict and the promise of an affordable unmanned reconnaissance and electronic warfare platform." The report observes, "RPVs will provide enhanced surveillance capabilities and, depending on the particular system, provide a capability to jam enemy communications, detect and destroy surface-to-air missile batteries, and deliver standoff munitions."[1]

The "new" capabilities outlined in the report overlay the uses of targeting and surveillance articulated in earlier drone experiments and went on to describe the drone within a field of electronic warfare. These developments can be tied to the technology's deployments in the Israeli-Arab conflicts that followed the Six-Day War in 1967. The RPV integrates surveillance with anti-aircraft defenses through a real-time image relay that uses television or infrared camera on the unmanned platform to transmit images of the battlefield to a nearby ground unit with a monitor. The image-transmissions system made it possible for operators to watch events on the battlefield in near-actual time. The relay capability of the RPV was, however, limited. The remotely piloted systems were operated from a ground-control unit based in a mobile trailer deployed behind the battlefield. Due to the size of the fuel tank onboard, the range between the unmanned platform and the receiver was around fifty kilometers and flight length was less than a few hours. The system's potential lay in its capacity to monitor military installations, mobile anti-aircraft units, and troop movements, *not yet* to attack individual militants.

The key judgments from the CIA report emphasize that RPVs, "when used as a reconnaissance system, may help prevent conflict and maintain stability in tense Middle Eastern and Asian areas" through "timely intelligence."[2] The report extends the logic of peacetime intelligence that had developed in the 1950s in the Cold War to monitoring ongoing conflicts via client states. The report also cautions, "A bomb-laden RPV provided to a terrorist group by a patron state could be used against a US embassy or other target in dramatic fashion. Although we have no indication that a Third World nation or terrorist group is planning such a modification, operators of RPVs can replace the surveillance equipment with high-explosive payload, effectively converting the RPV into a guided bomb capable of surprise attacks at short and medium ranges."[3] Later, the authors concede that the scenario they outline is unlikely; however, a terrorist attack by RPV is presented as a major finding nonetheless. The RPV could be theoretically converted to a platform for terrorist attack by turning the unmanned aircraft into a guided bomb. The U.S. military tested this possible use of the RPV decades earlier using the television-guided assault drone built in World War II, examined in chapter 2. By 1986, the formulation of unmanned aircraft as guided bomb was refashioned as a *terrorist* mode of attack, demonstrating shifts in understandings of the enemy from foreign enemy soldier to the trope of the suicide-bomber-as-terrorist. I emphasize how the U.S. military refashioned tactics described by the report as legitimate targeted killings, rather than terrorism, in the war on terror.

The CIA report does not merely indicate a historical ambivalence between the strategies used by the United States and tactics it labeled as "terrorist." Indeed, concerns raised by the 1986 report remain relevant today and received significant attention after the Islamic State used modified commercial drones to carry explosives in 2017. Instead, the report sets out a pattern whereby identification of possible threat becomes a strategy for attack; this interconnection sets out parameters for the field of war in which the drone aircraft is supposed to both act in defense against and attack. The CIA report identifies a sphere of threat via the RPV that articulates the drone as a potential vehicle for terrorism, imagining a field of threat that shapes and is shaped by the system itself. The RPV is organized by terrorism and the "Third World"; moreover, its very existence suggests the possibility of the proliferation of mini-RPVs in conflicts not defined by global superpowers, a subversion of the dominant military strategy of the time.

Although the report inventories remotely piloted vehicles produced in the United States, Canada, the Soviet Union, and other European countries, the writers emphasize the significance of Israel in RPV development. The report explains, "Israel is the leading RPV developer in the Third World and has more systems in production than any other nation."[4] By 1986, Israeli companies had sold drone aircraft to the United States, Switzerland, Singapore, and South Africa. Positioning Israel as both the leading developer of RPVs and a "Third World" country is significant. Labeling Israel as "Third World" in the report elides tensions that marked Israel's relationship with much of what was believed to constitute the "Third World," a term used to describe developing countries in Latin America, Africa, and Asia not aligned with the United States or the Soviet Union during the Cold War. For instance, it would have been used to describe Palestine, Lebanon, Syria, and Jordan, Israel's enemies in the 1980s. Another term in the 1980s often used to describe Israel, alongside South Africa, Taiwan, South Korea, and some Latin American countries, was "pariah state." This indicated its status outside the international order as "a small power with only marginal and tenuous control over its own fate, whose security dilemma cannot easily be solved by neutrality, nonalignment, or appeasement, and lacking dependable big-power support."[5] I point to this definition from the 1980s not to weigh in on its assessment but to suggest how the RPV in the CIA report fits with a field of war defined by perpetual insecurity on the margins of the international order.

This chapter considers how drone aircraft used by Israel for real-time reconnaissance in the 1980s were evaluated by the U.S. military. It looks at conflicts between Israel and Lebanon and Palestine through the interpretative lens of U.S. power to consider how military strategies conceived of as "Third World" became integral to post–Cold War conflicts. I organize the chapter around a historical punctum often overlooked in accounts of the rise of drone warfare. The acquisition of drone aircraft by the United States from Israel was part of what was, in 1991, the largest case of military corruption ever prosecuted by the Federal Bureau of Investigation (FBI). Former assistant secretary of the navy Melvyn Paisley and a network of associates were found guilty of accepting millions of dollars in bribes from the military industry in exchange for preferential contracts. This story was headline news in the United States and Israel. The extent of the corruption went well beyond the acquisition of the Pioneer RPV. The convictions followed widespread reporting on the mismanagement of unmanned aircraft programs in the United States. Between 1972 and 1987, the U.S. military spent close to one billion dollars on the development of a remotely piloted vehicle built by Lockheed known as Aquila, deemed a colossal failure.[6]

The previous chapters interrogated the failures interwoven with the human, machine, and media parts that make up the discursive formation that is the drone: innumerable crashes, image glitches, and navigational failures. These failures are papered over as part of the larger erasures and disavowals that make "unmanning" an ontology. The corruption scandal, like the crashes and equipment failures before it, destabilizes the emergent framework intertwining the RPV and the terrorist. In the process, "corruption" emerges as a structuring principle of the drone's interaction with the military-industrial complex, far exceeding the bounds of an individual scandal. "Corruption" is a term typically used to mark the failures of non-Western countries to achieve the ideals of liberal governance. In the 1980s, the term "corruption" would have been rife in characterizations of the "Third World." My use of the term does not aim to replicate this use. Rather, an analysis of the military procurement scandal in the United States undoes the supposed distance between First and Third World, making visible the extent to which bribery and military excess are overwritten by technical advance in the United States and are, through the overarching concept of "corruption," themselves etymologically and conceptually linked to the technical failures (crashes, navigation) of the drone. I draw here on the secondary meaning of corruption, "the process of causing errors to appear in a computer program or database."[7] Corruption highlights that the context that entangles the RPV, the "Third World," and terrorism is always already marked by error, an "unreadable" file. A drone's crash may be linked to its larger discursive formation, but the crash itself is particular: a single, clearly demarcated event. The error suggested by corruption, on the other hand, is both systemic and enmeshed with the organization of technoscience, state power in the United States, and the global order imagined by the creators of its weapon systems.

In *War and Cinema,* Paul Virilio, writing in 1989, draws a line from the emergence of military photography in the American Civil War to "today's use of video surveillance in the battlefield."[8] Virilio imagines that war at the end of the century is no longer defined by nuclear deterrence but "based on ubiquitous orbital vision of enemy territory. Rather like in a Western gun-duel, where firepower equilibrium is less important than reflex response, eyeshot will then finally get the better of gunshot. It will be an optical, or electro-optical confrontation."[9] He explains, "A war of pictures and sounds is replacing the war of objects (projectiles and missiles). In a technicians' version of an all-seeing Divinity, ever ruling out accident and surprise, the drive is on for a general system of illumination that will allow everything to be seen and known, at every moment and every place."[10] Virilio's analysis is a prescient account of shifts in the 1980s that led to a framework of warfare that emerged in the United States in the aftermath of the Cold War, emphasizing the significance of real-time image transmission for targeting, where eyeshot overlays gunshot (although concerns about nuclear attack remain as present today as they were at the end of the Cold War). Virilio's outline, however, mimics the hyperbole of military commanders and, in so doing, misses how practices of seeing and knowing are continually less perfect than they seem.

Against the line proposed by Virilio, I consider how the mini-RPVs developed by Israel were framed by the United States in the 1980s as making and responding to a new field of war. Rather than providing a ubiquitous, all-seeing view, the mini-RPV acts in the margins of the global order at the time—an order fashioned by insecurity, illegitimacy, and terrorist threat. Grainy, blurred images transmitted by RPVs, technical failures, and crashed drones all undo the omniscience Virilio proposes through electro-optical systems. One might rather think about what can be seen by real-time images as corrupted: evidence of systemic failures that undo the neat overlay of state power and eyesight to instead emphasize how the claim to being all-seeing is error-prone and ultimately undecipherable. Reading this history as a corrupted file instead calls attention to the limits and positions organized by unmanned aircraft, as well as the human and institutional frameworks that drive the projects.

Ryan Aeronautical's 124I and the Invention of the Modern Drone

In 1971, Israel approached Teledyne-Ryan, the manufacturer of the unmanned reconnaissance drone system used by the U.S. military in Vietnam, to purchase a model of the Firebee, the 124I, designed for the Israeli Air Force. Interviews, newspaper clippings, and magazine articles recount the sale of the drone reconnaissance system to Israel in 1971, though operational details of its use afterward remain classified. What is presented below comes from limited archival materials, which confirm that the 124I drones were used to collect visual and electronic intelligence over Egypt and Syria. The system remained in use by Israel until the mid-1990s and was also deployed for reconnaissance in the Lebanon War. It is likely that the extent of drone missions exceeds this scope.[11] The system was the first unmanned *reconnaissance* drone that Ryan Aeronautical sold internationally.

The sale, however, intentionally used the 124 model number to categorize the system as *target* drone. Although the exportation of Firebee targets to Japan for air defense training had been previously approved by the U.S. government, whether drone reconnaissance systems could be sold internationally was unclear. Their sale took advantage of the ambivalence between the target and reconnaissance systems, both based on the same airframe, to bypass possible export controls. The U.S. State Department granted approval for the sale of a target drone system to the Israeli Air Force, although it was openly known that this would not be the system's use. Robert Schwanhausser, in charge of the drone reconnaissance program at the time, explains, "Ray Ballweg was

Vice President of Teledyne Ryan in charge of our Washington office. He had been in on reconnaissance drones from the very beginning and handled the paper work through the State Department. He presented the whole story and openly explained that we were using the 124I nomenclature but that it was indeed a military reconnaissance vehicle."[12] Ballweg had also promoted the conversion of the Firebee from a target drone to a reconnaissance aircraft back in 1959, leading to the first secret project funded by the U.S. Air Force. Ryan Aeronautical's company history adds, "Because reconnaissance drones, being unmanned, are considered nonpolitical—and being unarmed are not attack vehicles—the State Department considers them defensive in nature rather than offensive."[13] This explanation by Ryan Aeronautical positions the drone as "nonpolitical" because it is unmanned and, moreover, not weaponized for attack. The drone's formulation as such extends the apparent distinction between technology and politics described in chapter 3 to the sale of military equipment internationally.

The sale of surveillance drones to Israel was enabled by the conflation of air defense training and reconnaissance, which was collapsed into a mutually beneficial miscategorization of the aircraft on the part of the Israeli government and Ryan-Teledyne. The 124I is listed as a target drone in Teledyne-Ryan Aeronautical's classification of unmanned aircraft.[14] In-house publications show that its payload was two film cameras that alternately collected low-altitude and high-altitude imagery.[15] This ambivalence in the technology of drone-as-target versus drone-as-reconnaissance-machine secured State Department approval for the sale. Moreover, the sale of the reconnaissance drone as a target system was justified by describing its use as nonpolitical. This logic generated international sales for Teledyne-Ryan, even as it disavowed the significance of reconnaissance in conflict and the interconnections between seeing and targeting inscribed in unmanned missions.

The 1973 Arab-Israeli War was a nineteen-day conflict between Israel and the surrounding Arab states, fought in Egypt and Syria. Israel used the 124I drones during the war to collect reconnaissance, particularly in well-defended areas like the Suez Canal. The neutrality claimed by the United States for unmanned systems would seem to be undone by the drone's role in this conflict, which suggests that, rather than a nonpolitical system, the platform could be utilized for both attack and defense. The aircraft's significance, however, would have been minimal. The 124I was susceptible to counterattack and technical failure. Of the twelve systems bought by Israel, only two remained in the aftermath of the war. Despite these failures, following the war, the Israeli Air Force journal *Heil Ha'avir* published an article about the aircraft, promoting its use as an alternative to piloted reconnaissance and the risks that conferred.[16] The Israeli military advanced unmanned aircraft as a possible way to counter improved air defense systems used by Egypt and Syria.[17] Early failures lay the groundwork for Israel's "invention" of unmanned aircraft in the following years.

In the 1970s, when Ryan Aeronautical attempted to publish an account of the sale of the Ryan Firebee drone to Israel, "Pentagon officials [who] read the manuscript . . . politely suggested that its publication be postponed indefinitely. By the time 'Security Review' would have finished censoring the story . . . there wouldn't be much left worth printing."[18] When the account cited here was published in 1982, efforts to build unmanned reconnaissance aircraft for the U.S. military were mostly defunded and largely seen as misguided. Benjamin Schemmer, author of the book's foreword, writes: "As this book comes off the press, not one U.S. remotely piloted vehicle is operational; indeed, the *only* 'RPV' [remotely piloted vehicle] that Teledyne Ryan Aeronautical now has in production is its original Firebee target drone."[19] This scenario is often cited as the starting point for contemporary unmanned reconnaissance, which points to the breakdown of the U.S. programs in the 1970s and notes how the next phase of unmanning draws heavily on the mini-RPV model developed in Israel beginning in 1974.

Mobile air defenses used by Egypt and Syria resulted in significant losses for the Israeli Air Force in the 1973 Arab-Israeli War, emphasizing the vulnerability of aircraft to ground attack. The mini-RPV was proposed as a response, which emphasized two qualities attributed to the RPVS: First, RPVs were tied to the protection of Israeli pilots. Second, the real-time images tracked mobile anti-aircraft defenses on the ground. As discussed in chapter 3, the capture of U-2 pilot Frances Gary Powers acted as a similar catalyst for unmanned reconnaissance in the United States. Yet, the use of the mini-RPV to track mobile air defenses positions pilot protection not within the broad project of "peacetime" intelligence but within the field of war. Mobile surface-to-air missile (SAM) units had transformed attack against aircraft. Battlefield reconnaissance might capture the location of a SAM in a photograph taken before an attack, but this position could be changed. The ability to monitor the battlefield in real-time through a television camera meant SAM units could be tracked by the RPV, relaying their actual location to commanders. The function of the RPV to counter a mobile target was thus first organized in a context that resembles a conventional field of war, even as it suggests how the system will later be used to track individuals.

Typically, the story of the Israeli "invention" of modern drone warfare begins in the aftermath of the 1973 Arab-Israeli War and uses the rationale outlined above as its basis. It leaves out the 1241 and instead focuses on a small-scale airframe conceived by Al Ellis, an Israeli-American aerospace engineer. This account interweaves Ellis's story with a broader, changing context of war, as well as through alliances and competing claims to the invention of the mini-RPV made by Israel and the United States. Repetitions and fault lines in the narrative complicate any single explanation, including the exigencies of the modern battlefield. Ellis left California in 1967 to work in support of the newly significant military industry in Israel, under sanction by other nations after seizing Jerusalem and the West Bank in the Six-Day War. He was a model airplane enthusiast and first built the mini-RPV as a hobby aircraft, along with three other devotees. His prototype, which became the basis for contemporary unmanned systems, was built from an airframe model supplied by Nick Ziroli, a hobbyist in upstate New York.[20] The aircraft had a wingspan of 12 feet and was operational up to 5,000 feet in the air, with an approximate airspeed of 100 miles per hour and a range of 60 miles. It could fly with a small camera on board. Ellis, who was employed by Israeli Aircraft Industries (IAI), approached IAI about the project but was passed over. Instead, Tadiran, an electronics company with extensive military contracts with the Israeli Army, supported the experiment.[21] The initial project led to the development of an RPV known as the Sorek, which was then purchased for the ground forces intelligence unit (the Nachshon) and, later, an updated model known as Mastiff. Before these projects were delivered to the Israeli Army, Ellis left Israel and returned to the United States.

Ellis's aircraft was not the only prototype tested by Israel. Teledyne-Ryan and Philco-Ford, two American companies, also secured contracts with the Israeli Air Force in 1973 for prototypes of an unmanned aircraft designed to track targets in real-time.[22] However, the two unmanned aircraft systems built in 1974 failed to meet the specifications of the Israeli Air Force, which was seeking a platform to support aircraft defenses against surface-to-air-missiles. Nonetheless, the Philco-Ford camera was later used in the mini-RPV prototype, and senior members of the U.S. military would lay claim to the mini-RPV's invention on this basis.[23] The mini-RPV was flown by radio control and operated from a ground-control car, which monitored the flight controls and received the televised image transmitted from the aircraft's camera. This arrangement, however, came together piecemeal, working from Ellis's airframe and the Philco-Ford camera to create the air vehicle. Experimental, composite materials eventually used to build the airframe were key to reducing the weight of the mini-RPV. Finding an engine that would work for the mini-aircraft and achieving takeoff and landing both proved difficult; in fact, the earliest systems were launched by catapult into the air.[24]

The development of the project at Tadiran led to a reconsideration of the mini-RPV. In 1978, IAI—the state-supported aircraft industry in competition with Tadiran—presented its experimental project known as the Scout (or the Zahavan in Hebrew); it was purchased by the Israeli Air Force. Internal competition between the two companies led to the use of rival mini-RPV systems by the intelligence forces and the air force from 1977 into the 1980s. The two projects were forced to merge in the aftermath of the Lebanon War. A company known as Mazlat, the Hebrew term for "unmanned aircraft," was created from the two projects. Mazlat sold a drone called the Pioneer to the U.S. Navy, leading to the corruption scandal discussed in this chapter.[25] In 1986, Tadiran was on the verge of bankruptcy and sold its share of Mazlat to IAI, which would go on to gain sole control of the venture. In Israel today, Tadiran is best known for manufacturing air conditioners. Israeli Aircraft Industries lays claim to developing the first mini-RPV, while the unmanned aircraft industry in Israel continues to be among the most significant worldwide.[26]

Meanwhile, during the 1980s, the U.S. Air Force pursued other strategies for weapons delivery against highly protected targets and aerial reconnaissance, including cruise missiles (of which the "American kamikaze" drone was seen as a precursor) and satellites. The U.S. military saw these systems as better suited to the operational goals of the time, which continued to focus on the Soviet Union as threat.[27] In 1982, the U.S. Army funded only one major project to develop unmanned aircraft: the Aquila. This project began as an experimental effort to create a surveillance platform that used television and laser guidance to track targets in real time. Lockheed developed the airframe, while Philco-Ford created the television camera system. The Aquila's development by the army reflects how mini-RPVs connected ground combat and air warfare. However, unattainable specifications for the project thwarted its development, and by the late 1980s, the drone was widely discounted as a colossal failure.[28] Rather than reading the innovation of the mini-RPV as a technological evolution, its development might rather be aligned with notions of a corrupted file. Ambivalence, misstarts, and competing claims to the drone undo the nonpolitical characteristics claimed for the program and underscore how the parts of the mini-RPV produce contradiction.

Lessons Learned in Bekaa Valley?

Internal Israeli military documents outline the contest between the IAI and Tadiran systems, used by different intelligence and air force units, even as tests in 1980 and 1981 evaluating the two models highlighted a reorganization of the battlefield by linking intelligence and targeting. One of the mini-RPV's aims (beyond the mere collection of intelligence, for which the mini-RPV had already been deployed) was to transmit battlefield information to a centralized command, integrating the distinct functions of intelligence gathering and target tracking. An internal report proposes, "Data from the field and data from the Forces, simultaneously with a 'real time' picture, will show the commander the situation of the Forces at any time."[29] Data relayed to "the commander" supposes a transparent picture of the war field in real time. Multiple accounts of what is often described as the first battlefield use of the mini-RPV by Israel in Bekaa Valley, Lebanon in 1982 show how that conflict became a turning point for renewed U.S. interest in unmanned reconnaissance while also complicating a transparent view.

Months after the initial evaluation, Israeli forces attacked the Bekaa Valley in Lebanon on June 9, 1982, in the first days of the First Lebanon War. Fighting continued until 1985. In two hours of intensive aerial bombardment on June 9, Israel took out the vast majority of the SAM defenses in Bekaa Valley, as well as numerous Syrian Air Force fighters called in to defend the SAMs. These hours of fighting were the "largest air battle since the Korean War," and Israel's losses were minimal. Rebecca Grant, writing in the June 2002 issue of *Air Force Magazine*, describes the battle: "The initial phase of that operation included a spectacular moment when the Israeli Air Force destroyed 19 surface-to-air missile batteries, with no losses, and downed a huge number of enemy aircraft. With *real-time intelligence* and careful exploitation of adversary weaknesses, the IAF dealt modern air defenses their first major defeat."[30] Her appraisal reiterates the assessment of many U.S. military analysts, who described the battle as paradigm-changing for modern war. I return to this battle, typically described as the first to deploy modern drone aircraft, to emphasize how the logic of "lessons learned" from battle shaped how mini-RPVs were conceived by the United States and the logic of drone interventions in war. These reports relied on a separation between Bekaa Valley and the rest of the First Lebanon War to construe the air battle as an unprecedented success. The reports mirror the discursive positioning of the drone as nonpolitical, emphasizing an unprecedented technological success against the much more morally fraught and contradictory outcomes of the war as well as ongoing violence in the region.

A *Time* magazine with the cover title "Israel Blitz," published on June 21, 1982, includes a report of the aerial fighting in Bekaa Valley under the headline, "Into the Wild Blue Electronically." It is worth contrasting with Grant's 2002 assessment, which emphasized the acquisition of real-time intelligence: *Time's* contemporary account instead describes it as an "electronic" battle. The article claims, "The successful Israeli air strikes on Syrian SA-6 mobile missiles in Lebanon's Bekaa Valley demonstrated that modern air war is as much a matter of computer array vs computer array as man against man." *Time* underscores the significance of American F-15s and F-16s flown by the Israeli Air Force, describing them as "two of the finest fighters in the world, flying electronic marvels." Innovations onboard F-15s and F-16s included the head-up display (HUD), "a mass of essential data" projected on the pilot's windshield, and computerized electronic countermeasure equipment (ECM). The report explains, "If a jet-propelled SA-6 were fired traveling toward the Israeli plane at 2,000 m.p.h., the jet's ECM would have singled it out for intense electronic jamming, trying to overcome the SAM's own anti-jamming system to send the missile veering off course." The article goes on to note the significance of Shrike radar-guided missiles, also built by the United States: "From as far away as 25 miles, the Shrikes' radar-seeking device can be tuned to the SAM's frequency—probably recorded by Israeli drones flown over the area before the strikes."[31]

The journalistic description above highlights how the air battle was described as electronic warfare in 1982. Rather than position real-time intelligence as central to winning the battle, the report instead suggests its relatively minor role in collecting radar signatures. I raise this point not to answer the question of which technology was the most significant to the Bekaa Valley air battle but to emphasize the multiple ways Israeli success has been contextualized. The perspective in the *Time* article reflects a U.S. bias and presupposes the preeminence of its fighter aircraft. At the outset of the war on terror in 2002, Grant returns to the air battle to emphasize the importance of unmanned reconnaissance aircraft. Her analysis reflects the emerging significance of unmanned aircraft for the United States, tying their use in Afghanistan and Yemen in the 2000s to the 1982 air battle.

Grant is not the first analyst to use the Bekaa Valley attack in this way. In *The Transformation of American Air Power*, Benjamin Lambeth outlines how the Bekaa Valley success became part of the repertoire of "lessons learned" for the U.S. Air Force, leading to what he describes as "the unprecedentedly effective performance of the allied air campaign against Iraq during Operation Desert Storm." Lambeth explains, "The impressiveness of that performance [in the Bekaa Valley] prompted a long-awaited U.S. Air Force fact-finding mission to Israel a year later . . . to 'go to school' on the Bekaa experience and ensure the right lessons were incorporated into the now much-improved American air combat repertoire."[32] The motif of "lessons learned" is found in much of the subsequent U.S. writing about the Bekaa Valley, which further instantiated its significance for modern warfare. It is used to bolster the supposed preeminence of air power in military conflict, which was challenged by advances to anti-aircraft gunnery and surface-to-air missiles.

In later writing, the brief mention of Israeli drones in the early *Time* magazine article is replaced by details that more fully outline their significance; notably, a number of these publications are from 2002, the year the U.S. military first executes Predator drone strikes in Afghanistan and Yemen. In "An Israeli Military Innovation: UAVs," Ralph Sanders argues that UAVs were inventions of the Israeli military industry. He explains, "Among the technologies that [Israel] has advanced are unmanned aerial vehicles (UAVs). Even though other nations have conducted experiments with these vehicles, Israel developed and fielded them as battlefield systems."[33] He provides details about the use of decoy drones in the Bekaa Valley battle, emphasizing their significance in trapping the SAM-6: "When the assault began, UAVs cruised the battlespace emitting dummy signals. Syrian radar operators thought that Israeli planes were attacking and launched most of their SAMs against unmanned vehicles. As the

Syrians reloaded and were vulnerable to air attack, Israeli fighters struck with telling effect."[34] Other commentators also highlight the significance of the new mini-RPV. Grant notes: "Israeli RPVs helped provide constant locations of Syrian SAM batteries."[35] She quotes the commander of the Israeli Air Force, David Ivry, providing his account of the use of the new system to track the mobile air defense systems: "'We tried to follow [the SAMs], because some of them had been mobile,' said Ivry. He added, that morning 'we'd been following them, all of them, [and] this was one of the conditions for that morning, to get all the information. Yes, we knew, no doubt, we knew all of them, where they were located.'"[36] Yet, these accounts by U.S. military analysts writing in 2002 may overstate two things: first, how the drone transformed command of the battle through television transmission and, second, the assertion that the success was a result of "new" technologies. The U.S. military analysts benefit from collapsing the multiple narratives of the battle of Bekaa Valley into a solely technological tale of success through the newly significant drone. There were instead multiple drone systems flown, including U.S. target drones, and the image transmission from the mini-RPV to the battle commanders may have been more limited than articles from the 2000s suggest. Analysts writing in the run-up to the war on terror read an idealized outcome of real-time intelligence transmitted by remotely piloted vehicles into the Bekaa Valley conflict, precisely because they are invested in the success of such air power in their own battles.

The Israeli Air Force web page "First UAV Squadron" briefly outlines its UAV program, beginning with the 1971 acquisition of the 124I from Ryan Aeronautical.[37] At the time, the unit also bought the Northrop Chukar, a target drone that was modified as a decoy. It mimicked the radar signature of a manned fighter plane. The web page notes that both systems were deployed during the First Lebanon War. In Bekaa Valley, two of the three Firebee systems were shot down, while the Chukars flew nine sorties and two failed during launching. Israel flew decoys before the actual fighter planes, attracting the surface-to-air-missile fire; the F-15s and F-16s would attack the gunneries after they had fired on the drone. The web page also highlights the significance of the newly tested IAI mini-RPV with a caveat, "On 11th June 1982 the Scout flight was deployed in the north and carried out the following missions: Collection of intelligence (in particular on AA [anti-aircraft] batteries), reporting the results of strikes, targeting of AA wagons and other armored vehicles, locating fallen aircraft and photographing trails so that they could be documented. During one of the operations an SA-8 AA battery was successfully identified and destroyed by a combat plane."[38] The website mirrors much of what is recounted later, but it sets the date of the Scout's usage to two days after the June 9 battle. There are various explanations for the discrepancy, including that the June 9 image-relay to the headquarters came from the Tadiran system; if so, it would have been operated by a separate unit based in Israeli intelligence and would not have been part of the air force squadron (or its history). Available archival materials from 1982 did not include documents from the Bekaa Valley battle. These discrepancies in detail, however, question the singular importance of the mini-RPV on June 9, 1982.

Writing in *Airpower Journal* in 1989, Matthew Hurley's "The Bekaa Valley Air Battle, June 1982: Lessons Mislearned?" perhaps unintentionally bolsters my account of the discursive iterations of the battle. He explains,

> Military analysts are always eager to derive "lessons" from recent military conflicts, but our perceptions of such lessons are often clouded by national biases; interservice rivalries; incomplete information; and differing needs, desires, and viewpoints. For example, the Bekaa Valley (Lebanon) air battle of June 1982 is widely regarded as a significant development in modern warfare. The Israeli Air Force (IAF) achieved a remarkable military victory, and certainly there are lessons to be learned from it. Unfortunately, most literature on the battle suffers from distortions resulting from the above factors.[39]

Hurley's article observes that the "lessons" taken from conflicts reflect not just the battle but the perceptions—tied to situated positions aligned with nation, military service, and desires—that also inform the interpretation. Accounts of the Bekaa Valley above are inflected by the viewpoint of the analyst rather than by registers of the battle. The significance attributed to the mini-RPV also comes from its later acquisition and use by the United States in the First Gulf War, as well as from the deployment of the Predator system beginning in 2002. The sweeping statements made regarding electronic war, real-time intelligence, and UAVs in Bekaa Valley are enmeshed with the contradictions between technology and politics enabled by drone aircraft.

Returning to the use of the mini-RPV in 1982, the Israeli Air Force web page makes clear that it was the information the system collected before the attack that was the most crucial. The Israeli military began monitoring Bekaa Valley in 1981 when the Syrian SAM installations were installed. "The [Zahavan's] first operational activity was carried out during the Lebanon missile crisis of 1981, when one of the UAVs was sent over the northern border and successfully broadcast real-time pictures of the Syrian antiaircraft systems deployed in the area."[40] This flight was followed by ongoing surveillance monitoring the installations, as well as the collection of signals intelligence and aerial photographs. Reports note that at least one drone system crashed in Lebanon during this time. These reports suggest that the mini-RPV's initial deployment began before the attack launched on June 9, 1982. The mini-RPV's significance lay not just in its ability to produce real-time information during the air battle but also in its role in Israel's preparations prior to the battle and in ongoing military actions by Israel in Lebanon.

When Israeli troops invaded Lebanon in 1982, Prime Minister Menachem Begin claimed that it was in response to terrorism against the Israeli ambassador to the United Kingdom, Shlomo Argov. Begin blamed Argov's attempted assassination on the Palestinian Liberation Organization (PLO), though it was carried out by a Palestinian splinter group led by Abu Nidal and assisted by Iraqi intelligence services. Begin indicated that Syria had provoked the air battle, while Israeli forces prepared to lay siege to Beirut. The same *Time* magazine issue that described the "electronic" battle in Bekaa Valley notes separately, "The attack, undertaken despite strong opposition from the Reagan administration, starkly revealed anew how little influence the U.S. has over its ally, Israel."[41] In its assessment of the Israeli invasion, the magazine states that the "portents were both uncertain and ominous." Importantly, while the battle was an opportunity to herald the successes of U.S. technology, the invasion itself and the conflict it began marked also the political and diplomatic limitations of the United States in its alliance with Israel.

A formerly classified July 1982 Israeli military report, "The Drone Operation for Ground Forces—Operation Shalom Hagali," evaluated the mini-RPV in the First Lebanon War. The document refers to the system's use from June 6 to June 25, 1982. It indicates that the two mini-RPV units, from intelligence and the air force, were integrated after the June 6 invasion. Summarizing its use over the time period that included the Bekaa Valley battle, the report states, "The systems unraveled at different localities (see Appendix A) and were designed to service/aid the Military Intelligence Directorate and the Northern Command/Intelligence and the Air Force commander and the North/Forward control post."[42] I did not receive the appendix in my document request and cannot verify actual instances. The report also explains that its evaluation was imperfect due to "difficulty in making cross-checks and verifications, difficulty reporting [from] the impressions made and deciphering the important from among the unimportant."[43] This is exacerbated in my case by the prevalence of redacted details. Nonetheless, the report conveys failure and tensions in relaying real-time information between the units operating the mini-RPV, ground forces, and headquarters—not the unqualified successes recounted in other reports and U.S. military assessments. The report notes "the lack of organization in coordinating and preparing sorties that were in immediate demand—and also with the demands that were received in the past for carrying out the next day—was felt."[44]

The challenges, however, did not minimize the potential of the mini-RPV. The investigation explains, "There's no doubt that giving intelligence in real time and during the duration of events would have allowed us to assess correctly, to command the forces correctly. . . . The drones' potential in this operation allowed this!"[45] Still, how video relay would be incorporated into military operations more broadly remained an open question. A summary of a discussion from December 5, 1982 concluded, "There must be many and diverse uses," but questions remained about how mini-RPVs and the two squadrons would be incorporated into the Israel Defense Forces (IDF) broadly.[46]

I quote these limited Israeli military materials to suggest that in 1982 what the drone was and how it would operate was not determined by a single battle at Bekaa Valley but rather continued to be worked out before and after the air battle. The mission catalyzed the mini-RPV program and emphasized its significance, but the dream of real-time intelligence relayed through a centralized command unit continued to be tested and was in an experimental phase of development. In hindsight, it is possible to read the battle as the inaugural use of the unmanned aircraft in modern warfare and as a portent of contemporary air power, but to do so overlooks the challenges and contingencies that were still being addressed at the time. Nonetheless, the possibility of remotely commanding war in real-time led Israel to call for an "emergency purchase" of new drone systems in 1982 and their continued development.[47]

Despite the spectacular qualities attributed to Bekaa Valley by military analysts, the results of the broader invasion and occupation of Lebanon were limited. Grant, whose unequivocally positive assessment of the air battle I quoted above, notes that the overall operation at best "produced mixed results."[48] Israel's official assessment is more direct: "The failure of Operation Peace of Galilee to achieve its objective prevailed upon the new national coalition government, which took office in 1984, to withdraw forthwith from Lebanon."[49]

Moreover, these qualified observations on the part of the U.S. and Israeli political and military establishments consider only the most narrow wartime definition of "success." Just as the atomic tests at Bikini Atoll were "successful" only if one ignores the displacement of the Marshallese and the contamination of the area with radiation, so too with the First Lebanon War. Indeed, the most conservative estimates suggest thousands of civilians died in Lebanon. Atrocities associated with the war include the massacres at Sabra and Shatila, carried out on September 18, 1982, by a militia connected to the Kataeb, a right-wing Christian party allied with Israel. The United Nations General Assembly condemned the massacre as a genocide.[50] While the pedagogy of "lessons learned" describes the unprecedented success of the Bekaa Valley air battle and its novel use of technology, the First Lebanon War itself was a fraught endeavor that had a far more ambivalent outcome.

Pioneer: The Drone as Corrupted Discourse

U.S. forces, which were on standby in Lebanon, evaluated the drone system in the aftermath of the Bekaa Valley battle, even as the United States held peace negotiations in the Middle East. In his memoir, *Fighting for Peace*, Caspar Weinberger, at the time the U.S. secretary of defense, describes his interest in the system in some detail. He recalls spending a day with his Israeli counterpart, Ariel Sharon. Together, they took a helicopter tour of Israel's settlements in the West Bank. Weinberger notes, "At the end of the day, I asked the then Defense Minister if it was all a 'coincidence' that every Israeli West Bank settlement I had seen occupied higher ground then [sic] any Arab settlement nearby. 'Of course not,' said Sharon, 'We have placed our settlements for the maximum military advantage.' " The next sentence in the book recalls, "I was also shown the Israeli camera-carrying drone, a remotely piloted vehicle that had made video tape recordings of me the day before, on my visit to the troops in Beirut." Weinberger explains, "It was a most impressive technical achievement," and goes on to claim that the system is actually a U.S. system, based on its resemblance with earlier mini-RPV prototypes built by U.S. military industry. Describing the motivations for acquiring the drone, he writes, "Later, I directed the Joint Chiefs to give us the same capability again: That Israeli drone had actually been developed by us but the Congress had refused to fund its deployment. It was then sold to the Israelis."[51]

The description of the camera-carrying drone comes immediately after Weinberger's helicopter tour of Israeli settlements, where he notes the military advantage of "higher ground." Weinberger thus intimates that the mini-RPV offers an analogous advantage, whereby the commander may achieve ongoing oversight of movements on the battlefield from higher—aerial—ground. Second, he describes the mini-RPV as a "technical achievement." Yet a third move reinscribes the technology as a product of the United States, claiming that it was actually "developed by us." A *Jerusalem Post* article responding to Weinberger's claim in 1992 notes, "For commercial reasons, there is a strong motivation to look for American innovation in non-American technology. The State Department in this way acts many times over as a patent lawyer for US companies looking to preserve or gain markets abroad."[52] Despite attempts on both sides to frame the initial technological exchange as apolitical, in the end, this blunt media assessment reinscribes the exchange and development within the framework of the state—and that state is itself marked by corruption.

Akhil Gupta's writing on the discourse of corruption in India sheds light on how corruption reveals the working of the state and its errors. He writes, "The discourse of corruption is central to our understanding of the relationship between the state and social groups precisely because it plays this dual role of enabling people to construct the state symbolically and to define themselves as citizens."[53] Gupta uses this analysis to trouble Western liberal ideals of the state and the divisions it presupposes. By turning this lens on corruption in the U.S. military in the 1980s, my reading of Gupta unsettles the usual geopolitical focus on corruption as a phenomenon limited to the developing world. Military and state excess and bribery underscore the imbrication of technology with politics, military institutions, and industry in the United

States. Moreover, like a faulty data file, the discourse of corruption—and by extension, the corrupted discourse—in the United States is forgotten, overwritten by an easily readable narrative of technological advance.

Although the major corruption scandal is connected to the acquisition of the Pioneer UAV, this system was not unique. The Aquila program developed by Lockheed for the U.S. Army was marred by overexpenditures, for example. In 1987, the U.S. Congress attempted to coordinate the various unmanned aerial vehicles in development at the time. The new oversight marked the first clear definition of the term "UAV," as well as an analysis of the more than one hundred systems that may have been precursors to UAVs. A chief motivation for this reorganization was the colossal failure of the Aquila program. The *New York Times* compared the Aquila with Israel's mini-RPV program in a March 28, 1988 editorial, "The Dumb Pursuit of a Smart Weapon." The article highlights the missteps in the development of RPVs in the aftermath of Vietnam, noting that projects cancelled by the air force were taken up instead by the army. The opinion piece explains, "The Air Force has never been interested in the dull, slow-flying planes needed to support ground troops. . . . Hence the Army was the only service interested in RPV's, and in 1974 began its Aquila program."[54] The program goals for the Aquila were complicated by the competing interests and bureaucratic practices that aimed to shape its development. Glossing over these problems, the *Times* explains, "Each bureau strives to add on new sensors, armor, frills and furbelows. The costlier a program gets, the more power to the program officer. The Aquila . . . is a prime example of this ruinous process. Development costs soared from $123 million in 1978 to nearly $1 billion by 1987. The unit price rose from $100,000 to a staggering $1.8 million." The editorial contrasts the Aquila with the system built in Israel: "Israel's Tadiran company took five years and $500,000 to make the Mastiff a superb weapon." The editors note that the Government Accountability Office (GAO) reported that "the Aquila proved hard to launch, regularly failed to detect its targets, and successfully completed only seven out of 105 flights." They observe, "After 14 years, it's such a disaster that even the Pentagon proposes to cancel it."[55]

The *Times*'s comparison of the two programs underscores the limits of the Aquila and the financial excesses of its development. Yet, the editorial may have too quickly seized on differences between it and the Israeli RPVs. That same year, news broke that the 100-million-dollar contract the navy awarded the Israeli company Mazlat for the Pioneer drone in 1985 had occurred in an "extraordinary manner." The impropriety involving a former top navy procurement official, Melvyn Paisley, was reported by the navy's project manager; he sent a memorandum explaining how the terms for the program were "dictated under 'a very innovative acquisition strategy' the purchase of an existing—'off-the-shelf'—pilotless aerial reconnaissance system." The memo included a document that "outline[d] some defects and workmanship problems with the systems." The memo also noted that nine of the forty-six drones purchased by the navy crashed.[56]

In 1991, Paisley pled guilty to federal conspiracy and bribery charges. Defense consultant William Gavlin pled guilty to bribery a year earlier, and implicated Paisley. "Galvin, 59, admitted that he, Assistant Navy Secretary Melvyn R. Paisley, and two Israeli businessmen conspired to influence the award of a Navy contract to guide remote-controlled drone aircraft."[57] The article explains, "As part of the deal, Galvin, Paisley, and the two Israeli businessmen were to split $2 million that an Israeli company, Mazlat, would place in a Swiss bank account, according to court papers." In turn, Paisley ordered the navy to purchase a ground-control station developed by Mazlat instead of pursuing a contract for the equipment from another company. Paisley pled guilty to three counts of bribery, conspiracy, and conversion of government property for accepting hundreds of thousands of dollars in bribes from navy contractors (extending well beyond the Mazlat scandal).

Just six years later, in 1997, the GAO provided an overview of the Pioneer program in its testimony about military UAV programs to Congress. The GAO makes no mention of the FBI corruption and bribery case against Paisley, explaining only that the Pioneer "skipped the traditional U.S. development phase of the acquisition process."[58] Its assessment of the project remains critical, however:

> The Pioneer began to encounter unanticipated problems almost immediately. Recovery aboard ship and electromagnetic interference from other ship systems were serious problems that led to a significant number of crashes. The Pioneer system also suffered from numerous other shortcomings. Ultimately, the Navy undertook a $50 million research and development effort to bring the nine Pioneer systems up to a level it described as a "minimum essential capability." Although Pioneer has never met objective requirements, the Navy and Marine Corps used the Pioneer in Operation Desert Storm, and operations in Somalia and, most recently, Bosnia.[59]

In line with this testimony, further reports on the unmanned aircraft in the late 1980s and early 1990s did not outline the technical prowess drone aircraft might possess or the strategic advantages they could offer for modern military warfare. Rather, articles announce the failures, setbacks, and significant limitations of UAVs, as well as the scandals marked by spending excess (in the case of the Aquila) and impropriety (in the case of Pioneer). Drone aircraft were headline news in every major U.S. newspaper between 1988 and 1991, which detailed the misuse of funds for their technological development and some of the spectacular functional inadequacies of the drone system.

Yet, the image of the all-seeing drone persisted. Corruption in the 1980s and its subsequent erasure in the 2000s accompany the rise of drone aircraft in the war on terror. Despite the negative press of the 1980s and 1990s, drone aircraft were continuously contextualized as *nonpolitical* technologies, beginning from the sale of the 124I to the American "lessons learned" in Bekaa Valley from the First Lebanon War. As an interpretive framework, corruption specifies the political and financial excesses that mark drone aircraft. Profit, individual, and institutional gains and power are organized in the name of the technological advance of military power. The "failure" of these projects is outlined not by mere error but in the ways that bribery and excess unsettle the "technical" results that are supposed to be achieved through unmanning. The corruption associated with this period of drone development allows us to trouble the terms by which the modern U.S. state is organized by the imperative of technological advance, which normalizes state power through the development of weaponry and attendant notions of enmity, conflict, and rationality. Yet, as this legacy of corruption demonstrates, these technological developments never advance with the normalcy attributed to them.

Targeted Killing: Eyesight as Gunsight from Bekaa Valley to Predator

The two frameworks that led to the development of separate mini-RPV systems in the late 1970s (namely, information collection for intelligence and real-time information for aerial targeting) came together to track and kill Hezbollah leader Sheik Abbas al-Musawi in what might be called the first instance of targeted killing by drone against an individual target. A *New York Times* article from 1992 explains, "Israeli forces killed the leader of the pro-Iranian Party of God in Lebanon today in a lightning strike by helicopter gunships that reportedly also left his wife, his son, and at least four bodyguards dead."[60] The strike was coordinated through a mini-RPV drone that followed al-Musawi's convoy, enabling the helicopter gunships to locate and kill the target. A publication commemorating the forty-year anniversary of Israel's UAV squadron recounts the attack in detail to emphasize the significance of the mini-RPV in carrying out the attack.[61] The missile system was not yet mounted on the drone aircraft but used television transmission to monitor an individual target and organize an attack in real-time. In the recollections reprinted in the commemorative pamphlet, Major Alon describes the moment of the strike from the perspective of the drone operators, who watch the attack in real-time:

16:09—A massive fireball followed by billowing smoke leaves no doubt, confirmed strike on the target. The rest of the vehicles in the convoy realize what happened and immediately stop.

An additional fired missile completes an exact strike and destroys the second vehicle. Doors fly open in the other vehicles and the occupants begin to flee.

We're there, watching every motion and conveying the intel to the air traffic control unit. In the image a vehicle was identified collecting the terrorists who were hit and then speeding towards Nabatieh. The order comes in, "Take him out." A few minutes pass by and at 16:32 the "Rokemet" formation (a second formation) destroys the fleeing vehicle with our full guidance throughout. The tension drops, we still do not know who was the target. In the briefing we were told it was a central and vitally important target. We couldn't conceive that we had just guided the attack that assassinated the secretary general of Hezbollah, Sheikh Abbas al-Musawi. At 17:00 there was already a news report, [when] we were already close to landing.[62]

Alon's memory of the event emphasizes the emotion of the operators, as well as recognition of their significance in a "vitally important" mission. Yet, even as the strike is described as an assassination, what stands out in his account is that the RPV operators do not know *who* they are targeting as they carry out the operation. The identity of the individual they are tracking onscreen is unknown until the strike is complete, indicating how the affect expressed by the operators is inscribed not in the images but in their framing.

After the strike, the Israeli government defended the assassination of al-Musawi by denouncing the party he led as a "murderous, terrorist organization."[63] Israel's Defense Minister, Moshe Arens, was quoted as saying: "It's a message to all terrorist organizations: Whoever opens an account with us will have the account closed by us."[64] Hanan Ashrawi, the spokeswoman for the Palestinian delegation to the peace talks, responded, noting the deaths of al-Musawi's wife and child: "To use the air force and state policy to kill women and children, that's not terrorism?"[65] At the time, the United States was trying to negotiate a peace deal between Jordan and Palestine and Israel. A State Department spokesperson issued the following statement: "We are concerned at the rising cycle of violence in the Middle East in recent days. We regret the loss of lives in Israel and Lebanon in recent days and urge all concerned to exercise maximum restraint."[66] These statements reflect what was at the time a U.S. policy against targeted killing.

This incident recalls the 1986 CIA report that contextualized the emerging use of mini-RPVs in the "Third World" through the lens of terrorism and insecurity. Israel justifies the strike against al-Musawi in 1992 as a response to terrorism, drawing on a longer history of assassinations secretly pursued by Israeli intelligence agencies.[67] In detailing the possibility of using a television-guided platform for terrorist attack, the CIA report never defines terrorism, though it does suggest that two Palestinian organizations, Fatah and the Popular Front for the Liberation of Palestine, might carry out such a strike. Reading the report in light of the 1992 attack outlines how the mini-RPV does not simply transmit a picture of the battlefield; rather, it organizes emerging conditions for war. In describing the possibility of using an unmanned platform for a terrorist attack, the report also suggests a counterattack.

Lisa Hajjar describes how the legal rationale for targeted killing emerged in the Second Intifada in Palestine in 2000: "Israel was the first state in the world to publicly proclaim the legality of 'preemptive targeted killing.' Officials asserted the lawfulness of this practice on the following bases: Palestinians were to blame for the hostilities, which constituted a war of terror against Israel; (2) the laws of war permit states to kill their enemies; (3) targeted individuals were 'ticking bombs' who had to be killed because they could not be arrested; and (4) killing terrorists by means of assassination is a legitimate form of national defense."[68] Hajjar outlines how the logic that "Palestinians were to blame" for the "war of terror against Israel" was used to justify both the deaths of individual leaders and the collateral deaths of civilians. These parameters are already suggested, if not yet articulated in law, in the 1992 strike against al-Musawi. Hezbollah terrorism was used to rationalize the strike as an act of legitimate war. Yet, the legal framework for targeted assassination highlighted by Hajjar is outlined in the logic behind the mini-RPV. The characterization of individuals as "ticking bombs" who cannot be arrested is conditioned by the electronic warfare described in contemporary accounts of the Bekaa Valley battle. The screen first used in the run-up to Bekaa Valley established a mode of seeing that layers mechanical feedback with the battlefield. The mission in 1992 extends and shifts this framework for drone aircraft to a counterterrorism measure.

A short clip of a drone strike created by the U.S. military and posted on the Department of Defense website demonstrates how the contemporary context of targeted killing draws on the political and military decisions that led to the use of the mini-RPV as a weapon in Israel. The video is an idealized portrayal of a missile strike guided by real-time image transmission where eyesight is gunsight. A title card before the video clip states, "Multi National Division Baghdad, MND-B, Soldiers fire missile from UAV, kill two terrorists, 3:30 a.m., April 9, 2008." Before any images appear, a text explains, "Soldiers using an unmanned aerial vehicle observed a group of terrorists with weapons attacking Iraqi Security and Coalition Forces with small arms fire in northeast Baghdad. The crew guiding the UAV fired one Hellfire missile and killed two of the armed terrorists."[69] The framework of the video and the image-transmission is defined by the state and its ability to identify and counter terrorism. This context is ostensibly created, however, by the terrorists—not the United States. The scene unfolds as a purported reaction rather than as an asymmetrical attack. The viewer takes on the perspective of the drone, where it is neither distance nor calculation that characterizes this position but threat and attack. The image is grainy, its infrared footage suggesting not the omniscience of sight but its blurriness.

In the initial seconds of the video, shots are fired from a gun by two figures who appear amid a row of parked cars and structures (figure 16). The shots create a scene that will at once resonate as one of terrorism, though it also recalls the interplay between aircraft and anti-aircraft gunnery. The strike is enacted in response to the threat below and organized as a chase scene. Onscreen, the drone detects the human bodies, identified by the title card as criminals, through their body heat. The camera follows the figures as they run toward the bottom of the screen. In the posted video, the initial gunshots captured by the image-relay are shown again after several seconds, this time zooming in on an individual who raises a weapon, gunfire bursting into the sky. The clip is edited to repeat the onscreen gunfire to frame the subsequent actions of the blotchy figures. The weapon's fire appears as black streaks against a monotone background, indicating a blurred threat.

The camera follows the shooters as they disappear around a corner and appear again, recaptured as the camera zooms in on the structures in the bottom part of the frame. The viewer then sees figures standing in an alleyway

FIG. 16 Video stills of MND-B soldiers killing two terrorists, April 9, 2008
Credit: Defense Video and Imagery Distribution System

behind a building. After a moment, they begin running again. Approaching the edge of the frame, there is a sudden explosion, inking over the image in black as fragments explode in the air. The Hellfire missile strike consumes the screen, and the image immediately cuts to black.

Even though the clip posits the drone as "soldier" and then opposes the "soldier" to the "terrorist," the Hellfire missile strike also establishes the technological advantage of the "soldier," as it responds to the small-arms fire by obliterating the human targets. Once the scene has been set up to portray a defensive counterattack, the scale of the missile strike overwhelms the chase scene that has come before. The inky black explosion insists that the "soldier" is literally on top, reaffirming the state power also inscribed in the image. The one-minute film stands in for how the Department of Defense would like to represent all unmanned combat air vehicles, though the footage represents only a minute of the hundreds of thousands of hours collected by drone aircraft since 2001. What if the picture were slowed down to capture the ongoing surveillance and its monotonous monochrome? The clip would instead be a brief blip in endless footage of mostly nonevents.

The clip is pedagogical, implicitly teaching the viewer how to see drone warfare. In this way, the image is always already corrupted by what it leaves out—namely, the broader context of war and the human actors that animate the image and the drone. The cat-and-mouse chase invites the viewer to watch the scene as if it were unfolding before her. Yet, this immersive engagement is produced by interactions between a pilot and sensor operator, networked through radio, digital, and visual communication. The video captured by the UAV can be watched by military personnel at the Pentagon or transmitted to forces on the ground. The scene is marked by crosshairs, as if only capture of the target mattered. The interplay of threat and attack in the video proposes a mechanical frame for warfare, disavowing the politics that both led to the scene and staged its aftermath.

Conclusion

● ●

Nobody's Perfect

Unmanning contends that drone aircraft perform politics through a disavowal of human action. The experiments analyzed from the interwar period to the end of the Cold War outline how this contradiction shapes contemporary practices of targeted killing, while these experiments' failures emphasize how the supposed negation of man is fashioned by the network of human, machine, and media parts. As a conclusion, I analyze a short piece of film propaganda, *Nobody's Perfect*, made to promote the sale of the Firebee, the target and reconnaissance drone developed by Ryan Aeronautical. Described as "the company's humorous movie, 'Nobody's Perfect' depicts some of the most laughable, though occasionally serious, test flights which went berserk,"[1] undoing the imagined ideal of unmanning as machinelike, dehumanizing and automatic. The scenes examined below draw out key themes of my argument: unmanning is paradoxically performed by humans; the mythos of technological innovation is tied to failure; and the humans, machines, and media that make drone aircraft establish conditions of violence through negation and denial.

Nobody's Perfect situates drone aircraft and their development in a history of breakdown. The film uses irony, crashes, and connections to persons to draw out the politics made and undone by the drone's parts, which are, in the end, attributed to "nobody." In Donna Haraway's theorization of the cyborg and the relations between human and machine that she extolls, she begins with irony. She writes, "Irony is about contradictions that do not resolve into larger wholes, even dialectically, about the tension of holding incompatible things together."[2] *Nobody's Perfect* reckons with a history of failures that are built into the drone and the tensions between human and machine that are enacted by the system. Failures in the film seem funny because the viewer thinks no human is harmed as the drones fall from the sky. Laughter instantiates the disavowal, refusing what is human about the assembled system. If pilots or crew died in the aircraft crashes, the scenes would be tragic. Instead, at the end, the company's engineers are characterized as learning to "bury their mistakes." Rather than promote unmanned aircraft as technological advance, the film situates the drone within a history of denial while burying its ties to the politics of disavowal emphasized in this book. As a counterpoint to what the military often touts as the utility of unmanned systems, the ironic film highlights their incoherence and breakdown, both of which haunt unmanning's ontology and current use.

The opening sequence of the film shows a succession of eccentric early aircraft and spectacular crashes. At the end of the montage, a Ryan Firebee drone explodes as it is launched. In the slowed-down shot, the careening drone flips through the sky as the aircraft bursts into flames and flies directly at the camera. The early aircraft that fail to fly are drawn from an Army Air Corps film, *Aeronautical Oddities*, a montage devoted to bizarre, experimental aircraft. *Nobody's Perfect* adds color footage from the Firebee crash to the sequence, and the title of the film appears over the billowing smoke from the explosion. The letter "r" is backwards in the title, referencing both the company's name and the imperfections touted by the film. As in this book, the film positions drone aircraft in a genealogy of oddities and failures. *Nobody's Perfect* counteracts the typical treatment of drone aircraft, which depicts their rise as a result of technological advance that replaces humans with machines. Instead, the film works from the overlap and connection between people and aircraft.

After featuring the Firebee drone in the opening sequence, the film turns to the role of Ryan Aeronautical in training Air Corps cadets. Humans acting as aircraft serves as the groundwork for the following scene and sets up the rest of the film. The voiceover announces: "During World War II, Ryan trained over 12,000 Air Corps cadets. Training was rigorous. Pre-flight was demanding. They were all very healthy . . . mentally." The beat of the music is interrupted in the next shot by the buzzing sound of an aircraft. In the image, a series of men run through the frame with their arms outstretched like wings just as the narrator says, "mentally" (figure 17). After they run through the screen, the voiceover pauses and continues, "Well, this is where they got their start." Pointedly, "the start" of the drone is a scene of humans performing flight. In the sequences, the interplay between human, machine, and media shows how humans enact what they are not and unmanning is paradoxically performed. The mechanistic qualities of the drone are illustrated through their enactment by humans and in the moments when the aircraft do not work as planned.

Moving forward in the film, the middle section is devoted to Ryan Aeronautical's Flexwing projects. These include both manned and unmanned versions

FIG. 17 Film still from *Nobody's Perfect,* Ryan air cadet training
Credit: San Diego Air and Space Museum

of a lightweight aircraft, which were forerunners of hang gliders, tested in the 1960s for the National Air and Space Administration (NASA). Rather than opposing human success and mechanical failure or vice versa, foibles in this section connect manned and unmanned flight. Drolly, the narrator remarks: "In 1962, our advanced engineering group conned management into thinking Flexwing was the vehicle of the future." The next shot shows the aircraft, dangling from a crane. "One of the more saleable features of this bird was the ease with which it could be assembled." The shot cuts to an engineer, who is underneath the billowing fabric of the Flexwing, while the narrator observes: "Anyone with a ten-ton crane and sixty helpers could have it ready to fly in a week's time under no wind conditions." The voiceover comments, "It was recognized throughout the industry that the project Pterodactyl set aviation back fifty years." The sequence pokes fun at innovation, as the Flexwing "con" never takes flight and the engineers struggle throughout the scene with the unwieldy aircraft.

The next scene introduces the first unmanned aircraft: "Headquarters: *Flexbee* Flats—Dead Man's Lake Division." Flexbee was a miniature, pilotless version of the Flexwing. Other than the name and size, however, little distinction between the two projects is made. Rather, it is the continuity of the triangular aircraft and its ongoing failure that bring the scenes together. Showing the Flexbee rolling across the desert, the sound shifts to that of a train as the system never achieves liftoff. In the next shot, the triumphal, military marching music returns and—as an apparent solution to the problem—the Flexbee appears on the roof of a military Jeep, launched through the momentum of the motor vehicle. A group of observers looks to the sky, and the camera follows the Flexbee's trajectory as it crashes into the desert. Five more crashes are shown in the scene, while the sequence concludes with a Flexwing falling from the sky with a human figure attached. The narrator jests: "As a result of his good work on the drone, the project engineer was allowed to test out our first furrowing parachute." The figure lands with a thud.

The sequence devoted to the Flexwing is replete with observers, engineers, technicians, and test pilots, pointing to how the experimental aircraft—manned or unmanned—rely on human support systems that extend beyond their operation in flight and perform their opposite. The scenes also make fun of human involvement and, in moments of dark humor, align the crashing aircraft with human risk and death. As opposed to the official photographs and films described in earlier chapters, *Nobody's Perfect* emphasizes how humans and technologies are entangled. Here, what was previously unmanned is, rather, collectives of persons and aircraft systems that constitute the projects. The character of the engineer is literally attached to the drone in the scene of the furrowing parachute. Drone crashes counter the apparent self-propelled technical evolution, emphasizing the role of human actions in its development and their "con" that leads to further funding.

Next, the film shows a guided missile being tested against a Firebee target. Again, the shift is unexplained, moving smoothly from experimental aircraft to target drone. In the opening shot of the scene, a uniformed soldier is on the telephone. The narrator explains: "The Army calls for a missile to kill the *Firebee*." The background music is a waltz, and the launched missile splits in half as it shoots into the sky. Continuing, the voiceover explains: "And this is what happens when one section of engineering works on thrust and the other section works on guidance." The missile comes apart in two pieces and circles through the sky in an elaborate dance. With the words, "At MacGregor base they claim a high number of flights per target. But with missiles like this, who can lose? Boy, that Firebee is an elusive target," the missile comes crashing into the desert. Yet, failure is also shared by the Firebee. The narrator proposes, "Let's try again," returning to the army soldier on the telephone. In the next shot, the Firebee is launched and crashes, producing a large explosion. In the scene, all of the parts of the war simulation are broken. Rather than a theater of attack and defense, both sides fail, as the missile guidance falls apart and the drone crashes after launching.

In the following scene, a black cloud of smoke shoots up from the air, and the film cuts to an aerial view of the crash. This part ties mechanical failure to death—an ironic response to the violence made possible by missiles and drones.

FIG. 18 Film still from *Nobody's Perfect,* mourners at drone crash
Credit: San Diego Air and Space Museum

Three men dressed in black walk over to inspect a crashed Firebee. "We kept a professional staff of undertakers, mourners, and rock kickers on the payroll," explains the voiceover. They pick up the parts of the Firebee strewn over the crash site, in the first of several scenes that highlight the pieces of the crashed drone (figure 18). The men in the scene seem to pay homage to the crashes in the previous sequences, attending to the scattered parts. The "life" and "death" of the unmanned aircraft play out an uncanny reversal of human and machine, mourning for the drone even as the scenes suggest the violence that can be carried out through the weapons. The scene is a sardonic and unintended prelude to funerals that mark the use of the Predator drone in the war on terror. In 2012 alone, the Bureau of Investigative Journalism reported twelve independently confirmed attacks against rescuers and mourners by unmanned aircraft in Pakistan during funerals.[3] In this uncanny resemblance, death haunts the ironic tone of the film and recalls the disavowed consequences of late modern warfare and aerial attack. The middle part of the film links death and breakdown to illustrate how the performance of machine-like power is haunted by the disavowal of colonial and racial violence that is recounted in *Unmanning*.

The next segment of the film situates the drone within the proxy conflicts of the Cold War and ties the Firebee's use to national protection against threat. The scene inadvertently foreshadows the drone's use for counterterrorism through the drone's ironic hijacking. The sequence begins with domestic tranquility: three men in short-sleeve shirts and trousers lie on a small knoll of grass. The voiceover notes, "At Tyndall Air Force Base, Ryan has established an outstanding crew. The leisure and relaxed atmosphere depicted here is a result of their ability to turn a bird around faster than any other base in the world." Calm string music is in the background. The music crescendos as the shot pans across to the drone and cuts to the control room. Changing tone, the narrator observes, "However, with their proximity to Cuba, they do have their little problems." A man enters the control room with gun and a handwritten sign printed on the back of a manila folder: "Take this drone to Havana." A close-up of the demand is followed by a close-up of the hijacker. He wears military fatigues and sunglasses, smokes a cigar and has a mustache—a caricature of a Cuban revolutionary and stereotype of a Latin American. The next shot shows the operator's hand shifting the controls. Showing the analog display charting the drone's course, the needle that marks the location of the Firebee moves sharply to the south.

"Take this plane to Cuba" became a refrain for aircraft hijackings in the 1960s. Between 1968 and 1972, there were 137 hijacking attempts in the United States; 91 of these were attempts to take aircraft to Cuba. Robert Holden writes, "The hijackings to Cuba during that period were largely routine. Airliners carried approach plans for the Havana airport and crews were instructed not to resist hijackers. There were also standard diplomatic procedures for obtaining the return of planes and passengers."[4] "Take this plane to Cuba" was even picked up in comedy sketches of the period. The hijacking crisis led to the installation of metal detectors in airports, setting the groundwork for contemporary security measures. Inscribing the Firebee in this scenario suggests that the drone must be protected from threat, tying the previous scene of mourning the machine to its position in national defense. The drone is set against a stereotyped enemy from the Cold War typified by racialized tropes of Cuban hijackers and militant actions that play out in the quotidian. Yet, the scene is also marked by the disconnect between the operators and hijackers in the control room and the drone's careening course on the chart. It is as if the performance between threat and its response never actually cohere. Rather, the scene highlights how this interplay is, rather, performed as propaganda for Ryan Aeronautical.

The last sequence of the film repeats the earlier shot of the exploding Firebee on a launch pad, returning to the role of the drone as a simulated enemy that piloted aircraft attack. The narrator says, "And here a fearless fighter pilot rushes to an alert." After showing the launch pad with the drone, the scene moves to a fighter pilot jogging calmly to the aircraft on the runway; the next shot is sped up to make it look as though he is rushing to the cockpit of a fighter jet. It seems as though the final scene will set up the opposition between human and machine that often characterizes portrayals of the drone. Yet, this does not happen. Cutting back to the drone on the launch pad, the film uses a feminized voice for the first time. In a husky tone, she counts down, "4–3–2–1–Blast off—hah." Her voice sexualizes the launch of the drone system; the feminine voice invites the viewer to imagine the scene as a climax. The drone whirls through the air as it bursts into flames. As metal parts are strewn through the air, the drone crashes behind a large outcropping of bushes in front of a nearby building. In the end, it is the failure of unmanning that is portrayed, emphasized by the feminine voice that announces the connected breakdown between human and machine.

The film ends with the foibles of men, as opposed to a climactic triumph. Ryan Aeronautical highlights failure, not the ability of the drone to dominate from the air, to mark the system's "success." This paradox contests the ideal of the inhuman drone that is made to displace human action. The music ends, and the voiceover explains, "If we have learned nothing throughout the history of our developmental work, our engineering department has learned very well how to bury their mistakes." On a beach, in a sped-up shot of a man shoveling, a pile of dirt rises on the screen. In the next shot, Richard Strauss's *Thus Spoke Zarathustra* provides the background music for a slow panning shot of the debris from the drone as the closing credits roll (figure 19). This is the same iconic music from the 1968 film *2001: A Space Odyssey*. In *2001*, the music accompanies the triumph of the ape with the bone tool that he uses to violently kill his prey. Violence, in *Nobody's Perfect*, is rendered through irony. The repeated failures and numerous drone crashes do not build a seamless connection between human and machine, even as they become more entangled. In the final scenes, there are merely the strewn and broken parts of the drone, which have become buried by the present. Human and nonhuman are linked by contradiction and negation.

In the war on terror, targeted killing is framed by the U.S. government as maximizing national protections through technological advance while absolutely minimizing loss of personnel in a global sphere of permanent threat and attack. *Nobody's Perfect* indicates how this story is built on a buried history that defies this characterization and, rather, suggests the ongoing presence of human action and wartime violence that exceeds calculated utility. Hugh Gusterson notes that the U.S. drone is narrated as a "technology that is almost magical giv[ing] its owners, who are looking on the scene from high in the sky, a godlike power over life or death."[5] "Godlike power" is pitted against persons targeted on the ground through real-time video and an integrated, laser-guided missile. The scenes build asymmetry, describing U.S. protections as invulnerability, while a target is tracked and killed. They propose a calculus that weighs the ideal of a more-than-human national defense and an absence of risk for U.S. personnel against threat, suggesting unmanning's indispensability.

FIG. 19 Film still from *Nobody's Perfect,* debris from drone crash
Credit: San Diego Air and Space Museum

Human, machine, and media parts, however, produce unmanning in ways that disrupt the technological determinism and calculus claimed by the U.S. government as its basis. Drone strikes do not actually negate human action, nor are they the result of mere technology. Rather, they continue to be animated by human operators and the policies of U.S. governments. The weighing of defense and attack is tied with a particular scene of war and its political basis. The aircraft are not all-seeing; rather, a drone point-of-view reiterates the paradoxical authority of the U.S. government to both define and counter terrorism, in the case of today's drone strikes. The designation of threat is performative. It raises the questions: How is the "terrorist" made by the drone? What is disavowed in the determinations made by the U.S. president or military by describing targeted killings as morally and politically necessary U.S. protections? How is this claim circumscribed by the ostensibly inhuman drone? Machinelike invulnerability is disrupted by unmanned aircraft as an interpretative frame—also subject to breakdown—where determinations imply multiple possible and debated outcomes rather than moral and political necessity.

Unmanning unsettles the appeal to technological obduracy made by contemporary advocates and critics, complicating claims of technological invulnerability as political right. Drone failure is a lens to rethink unmanning, undoing its negation of politics for technology through a double negative, wherein the human reemerges. And failures persist in contemporary systems. Between 2001 and 2015, more than 237 military drone crashes either destroyed or did more than two million dollars' worth of damage to U.S. unmanned aircraft, representing approximately one-third of the fleet.[6] The crash rate for air force drones has remained steady between 2011 and 2017, averaging thirty-nine unmanned aircraft per year.[7] Broken airframes are a counterpoint to scenes of unmanning as an all-seeing eye, instead pointing to the erasures necessary to imagine the drone as merely a technical frame for war and to the limits of human action denied. The paradoxical performance of drone warfare by humans, media, and machines establishes a disavowal of the politics of drone use; *Unmanning* contests the drone's mythos by emphasizing how human action and failure underwrite the negation it names.

Acknowledgments

When I began writing about drones in 2009, unmanned aircraft had not yet emerged as a major topic of public discussion. My dissertation committee recognized the potential of this project. I am grateful they saw its relevance and assured me, at some point, others would as well. Throughout my graduate studies, David Bates has been a thoughtful mentor, encouraging adviser, and critical reader and is an exemplary scholar. Charis Thompson's work in science and technology studies was an early model for me, and I have been delighted to think with her during the past several years. Jake Kosek has shared my interest in drones (by way of bees), and I am thankful for his engagement with my research. He is a challenging reader and enthusiastic interlocutor. Samera Esmeir's attention to my arguments and writing provided welcome advice, guiding me to bring forward my voice and foreground the significance of my ideas. Each one has been an inspiration for my scholarship.

In the Culture and Politics Program at the Walsh School of Foreign Service, I have found a home for my interdisciplinary scholarship through the program's unique approach to the study of power, politics, and culture. Shiloh Krupar is the best colleague I could ever hope for, and the support and comradery of the other CULPies, Katrin Sieg and Maria Luise Wagner, is exceptional. In the school, Fida Adely, Rochelle Davis, Lahra Smith, Mark Giordano, Scott Taylor, Shareen Joshi, John Tutino, Erick Langer, Daniel Byman, Emily Mendenhall, Sarah Johnson, Marwa Daoudy, Diana Kim, Toshihiro Higuchi, Carol Benedict, Charles King, and Kate McNamara all provided support and feedback. The faculty chairs, Jeff Anderson and Irfan Nooruddin, have been advocates of my work. In the college, I am grateful to ongoing conversations with Mubbashir Rizvi, Nicole Rizzuto, Caitlen Benson-Allott, Nathan Hensley, Brian Hochman, Denise Brennan, Marden Nichols, Ricardo Ortiz, Meg

Jones, and Mathew Tinkcom. I presented a version of chapter 2 to the America Initiative and benefitted from the group's thoughtful reading. My research was supported by generous funding from Georgetown University, including a grant from the Mortara Center and the Office for Global Engagement to support a manuscript workshop.

The intellectual community in the Department of Rhetoric and at the University of California, Berkeley, fostered this project in its early stages. In 2009, Paul Chan, after a guest lecture at the Berkeley Art Museum, asked if anyone was interested in filming an intervention outside of Las Vegas about drone warfare. Claudia Salamanca and I volunteered. This trip changed what I planned to do in graduate school and led to this project. At Berkeley, I benefitted from the companionship of numerous scholars and graduate students. The members of my cohort, Matt Bonal, Amanda Armstrong, and Paul Nadal, share a commitment to politically engaged, critical thinking, and I learned from each of them. Juliana Chow and Maude Emerson helped organize the Mediating Natures working group, which was formative in my intellectual development and a highlight of graduate school. Aglaya Glebova was an interlocutor throughout the dissertation, and I have relied heavily on and benefitted a great deal from her feedback. I am also thankful to Emily Carpenter, IK Udekwu, Suzanne Li Puma, Simona Schneider, Amira Silmi, Javier Arbona, Nick Sower, Gillian Osborne, Keerthi Potluri, Kris Fallon, Alenda Chang, Rima Hussein, Anjali Nath, Kelli Moore, Margaret Rhee, and Milad Odabaei for their thoughtful engagement with my work. At the Berkeley Center for New Media, Gail De Kosnik and Ken Goldberg both encouraged my research and helped me to think unmanned aircraft as new media. I am also grateful to Eric Paulos, who organized the Robots and New Media Conference and included me in the process. At the Berkeley Center for Science, Technology, Medicine, and Society, Sam Evans and Cori Hayden both encouraged this project. Mireille Rosello and Margriet Schavemaker were early supporters of my research at the University of Amsterdam.

I received fellowships for my dissertation research from the Townsend Center for the Humanities and the Berkeley Center for New Media, which were critical for both the research and writing phases of this project. The Townsend Center for the Humanities 2013–2014 Fellows provided feedback. I am thankful to the Cultures of Surveillance Conference at the University College of London, 2011; the Barcelona Centre of International Affairs, 2011; the Gender, Bodies, Technology Conference at Virginia Tech, 2012; Speculative Visions of Race, Science, Technology, and Survival at the University of California, Berkeley, 2013; the Princeton-Weimar New Media Summer School, 2013; the Society for Social Studies of Science Conference in San Diego, California, 2013; Modes of Techno-Scientific Knowledge Workshop, 2014; the Center for Race and Gender at the University of California, Berkeley, 2014; the Pan-Optics

Conference at the University of California, Berkeley, 2014; the Society for Cinema and Media Studies Conference in Montreal, 2015; the Society for Social Studies of Science, 2015. Senior scholars at these presentations, including Lucy Suchman, Caren Kaplan, Lisa Parks, and Hugh Gusterson, all offered insight into the project.

At the National Air and Space Museum, Roger Conner, curator of unmanned aircraft, was enormously helpful in aiding my research and I benefitted from my conversations with Richard Whittle as well. I am thankful to Elizabeth Borja and Brian Nicklas for their guidance in the National Air and Space Museum Archives. Alan Renga at the San Diego Air and Space Museum Archives was also enormously helpful. I am grateful to Yehuda Chen and Tali Rotschild for their assistance in gaining access to materials from the Israel Defense Forces Archives, as well as Nitzan Gabai, Sara Grayson, and Dafna Ruppin for their assistance translating documents. Karen Rice, Tara Mendola, Andrea Miller, and Connie Yang all provided thoughtful editorial assistance in the final stages of the manuscript development.

Many thanks to all my friends who fill me with wonder and joy. I'm so glad I know you and we find each other, here and there, on this planet. To my family, I am grateful for their love, curiosity, life-long learning, propensity for debate, and belief in the greater good they have shared with me.

Notes

Introduction. A "Different Lethality"

1 See "Afghanistan/War on Terror/Predator Attack," *CBS Evening News*, February 6, 2002, Vanderbilt Television News Archive. http://tvnews.vanderbilt.edu/diglib-fulldisplay.pl?SID=20140427532989501&code=tvn&RC=653235&Row=4; "Afghanistan/Al Qaeda/Missile Strike," *CNN Evening News*, February 6, 2002, Vanderbilt Television News Archive, http://tvnews.vanderbilt.edu/diglib-fulldisplay.pl?SID=20140427532989501&code=tvn&RC=656391&Row=4; "Afghanistan/War on Terror/Missile Attack," *NBC Evening News*, February 7, 2002, Vanderbilt Television News Archive, http://tvnews.vanderbilt.edu/diglib-fulldisplay.pl?SID=20140427532989501&code=tvn&RC=654761&Row=9; "America Fights Back/Afghanistan/Bin Laden Hunt/Missile Attack." *CBS Evening News*, February 8, 2002. Vanderbilt Television News Archive. http://tvnews.vanderbilt.edu/diglib-fulldisplay.pl?SID=20140427532989501&code=tvn&RC=653272&Row=13.

2 Thom Shanker and James Risen, "A Nation Challenged: Raid's Aftermath; U.S. Troops Search for Clues to Victims of Missile Strike," *New York Times*, February 11, 2002, https://www.nytimes.com/2001/09/11/national/a-nation-challenged.html.

3 Qtd. in ibid.

4 See also my account of this strike in Katherine Chandler, "5,000 Feet Is the Best: Re-viewing the Politics of Unmanned Aircraft Systems." *Knowledge Politics and Intercultural Dynamics* 5, no. 1 (2012): 63–74.

5 Shanker and Risen, "A Nation Challenged," my emphasis.

6 Grégoire Chamayou, *A Theory of the Drone*, trans. Janet Lloyd (New York: New Press, 2015), 66.

7 Chris Woods, "The Story of America's Very First Drone Strike," *The Atlantic*, May 30, 2015, https://www.theatlantic.com/international/archive/2015/05/america-first-drone-strike-afghanistan/394463/.

8 Ibid.

9 "Creech Air Force Base—Fact Sheet," *Nellis Air Force Base* (blog), July 12, 2012, http://www.nellis.af.mil/About/Fact-Sheets/Display/Article/284172/creech-air-force-base/.

10   Qtd. in Esther Schrader, "Response to Terror; Military to Probe Alleged Abuse of Afghan Prisoners; War: Rumsfeld Orders an Inquiry into Reports That U.S. Troops Beat Villagers Taken in Raid. Meanwhile, Officials Deny CIA Missile Strike Killed Civilians," *Los Angeles Times*, February 12, 2002, http://search.proquest.com/docview/421724415?accountid=14496.

11   Qtd. in Rowan Scarborough, "U.S. Strike Likely Killed Top Terrorist; Finance Official Thought Dead," *Washington Times*, February 13, 2002, my emphasis.

12   Doug Struck, "Casualties of U.S. Miscalculations; Afghan Victims of CIA Missile Strike Described as Peasants, Not Al Qaeda," *Washington Post*, February 11, 2002.

13   James Dao, "Raid's Aftermath," *New York Times*, February 12, 2002, http://search.proquest.com/docview/92200997?accountid=14496.

14   John F. Burns, "A Nation Challenged: The Manhunt: U.S. Leapt before Looking, Angry Villagers Say," *New York Times*, February 17, 2002, https://www.nytimes.com/2002/02/17/world/a-nation-challenged-the-manhunt-us-leapt-before-looking-angry-villagers-say.html.

15   Qtd. in Vernon Loeb, "Alleged Beating of Prisoners Sparks Inquiry; Military Defends February 4 Missile Strike." *Washington Post*, February 12, 2002.

16   Derek Gregory, "From a View to a Kill: Drones and Late Modern War," *Theory, Culture & Society* 28, no. 7–8 (2011): 193. He revises his assessment in "The Territory of the Screen," *Media Tropes* 6, no. 2 (2016), https://mediatropes.com/index.php/Mediatropes/article/view/28054.

17   Burns, "A Nation Challenged."

18   As Chris Woods explains, the first Predator strike instead happened on October 7, 2001 in Afghanistan and was a botched attempt to kill Taliban Mullah Mohammad Omar, who escaped. See Woods, "The Story of America's Very First Drone Strike." My research, however, works against the idea of a "first drone strike," and, instead, I show its variable construction. What interests me is what discursively becomes the first drone strike, which is the November 3, 2002 attack in Yemen, the political stakes of this formulation, and the erasures proposed by it. I begin with the February 4, 2002 strike because it is the "first" widely reported in the U.S. media.

19   James Risen and Judith Miller, "Threats and Responses: Hunt for Suspects; C.I.A. Is Reported to Kill a Leader of Qaeda in Yemen," *New York Times*, November 5, 2002, https://www.nytimes.com/2002/11/05/world/threats-responses-hunt-for-suspects-cia-reported-kill-leader-qaeda-yemen.html.

20   David Johnston and David E. Sanger, "Threats and Responses: Hunt for Suspects; Fatal Strike in Yemen Was Based on Rules Set Out by Bush," *New York Times*, November 6, 2002, https://www.nytimes.com/2002/11/06/world/threats-responses-hunt-for-suspects-fatal-strike-yemen-was-based-rules-set-bush.html.

21   James Risen, "Threats and Responses: Hunt for Suspects; An American Was among 6 Killed by U.S., Yemenis Say," *New York Times*, November 8, 2002, https://www.nytimes.com/2002/11/08/world/threats-responses-drone-attack-american-was-among-6-killed-us-yemenis-say.html.

22   Julian Borger, "US Air Wars under Trump: Increasingly Indiscriminate, Increasingly Opaque," *The Guardian*, January 23, 2018, https://www.theguardian.com/us-news/2018/jan/23/us-air-wars-trump.

23   Donna Haraway, "A Cyborg Manifesto," in *Simians, Cyborgs, and Natures: The Reinvention of Nature* (New York: Routledge, 1991); Bruno Latour, *We Have Never Been Modern* (Cambridge, MA: Harvard University Press, 1993).

24 Sheila Jasanoff, "Introduction" to *States of Knowledge: The Co-production of Science and the Social Order*, ed. Sheila Jasanoff (London: Routledge, 2004), 2–3.

25 Charis Thompson, *Making Parents: The Ontological Choreography of Reproductive Technologies* (Cambridge, MA: MIT Press, 2005), 8.

26 Sigmund Freud, "The Infantile Genital Organization of the Libido [1923]," in *Collected Papers,* vol. 2, trans. Joan Riviere (New York: Basic Books, 1959).

27 Judith Butler and Biddy Martin, "Cross-Identifications," *Diacritics* 24, no. 2/3 (1994): 3, https://doi.org/10.2307/465160.

28 While recent scholarship has addressed "unmanning," gender, and sexuality, this book departs from this scholarship to examine the performative and political contexts through which unmanning as a concept emerges. For example, see Cara Daggett, "Drone Disorientations: How 'Unmanned' Weapons Queer the Experience of Killing in War," *International Feminist Journal of Politics* 17, no. 3 (2015): 361–379; Lauren Wilcox, "Embodying Algorithmic War: Gender, Race, and the Posthuman in Drone Warfare," *Security Dialogue* 48, no. 1 (February 2017): 11–28; Mary Manjikian, "Becoming Unmanned: The Gendering of Lethal Autonomous Warfare Technology," *International Feminist Journal of Politics* 16, no. 1 (2013): 48–65; Lorraine Bayard de Volo, "Unmanned? Gender Recalibrations and the Rise of Drone Warfare," *Politics & Gender* 12, no. 1 (2016): 50–77.

29 Ian G. R. Shaw, "Predator Empire: The Geopolitics of US Drone Warfare," *Geopolitics* 18, no. 3 (July 2013): 554.

30 Achille Mbembe, "Necropolitics," trans. Libby Meintjes, *Public Culture* 15, no. 1 (2003): 12.

31 Frantz Fanon, *The Wretched of the Earth*, trans. Richard Philcox (New York: Grove Press, 2004), 7.

32 Avery Gordon, *Ghostly Matters: Haunting and the Sociological Imagination* (Minneapolis: University of Minnesota Press, 2008), 25.

33 Katharine Hall Kindervater, "The Emergence of Lethal Surveillance: Watching and Killing in the History of Drone Technology," *Security Dialogue* 47, no. 3 (June 2016): 223.

34 Marshall McLuhan, *Understanding Media: The Extensions of Man* (Cambridge, MA: MIT Press, 1994), 8.

35 Friedrich A. Kittler, *Gramophone, Film, Typewriter* (Stanford, CA: Stanford University Press, 1999), xxxix.

36 Ibid., xl.

37 Donna Haraway, "Situated Knowledges: The Science Question in Feminism and the Privilege of Partial Perspective," *Feminist Studies* 14, no. 3 (1988): 587.

38 Jeremy Packer and Josh Reeves, "Romancing the Drone: Military Desire and Anthropophobia from SAGE to Swarm," *Canadian Journal of Communication* 38, no. 3 (September 14, 2013).

39 Drone operators, for example, typically describe a drone view like looking through a straw. See Thomas E. Ricks, "5 Big Problems with the Drone Programs." *Foreign Policy—Best Defense* (blog), December 10, 2015. https://foreignpolicy.com/2015/12/10/5-big-problems-with-the-drone-programs/; Richard Whittle, *Predator: The Secret Origins of the Drone Revolution* (New York: Henry Holt and Company, 2014), 131; Hugh Gusterson, *Drone: Remote Control Warfare* (Cambridge, MA: MIT Press, 2016), 31.

40 Lisa Parks, "Drones, Vertical Mediation, and the Targeted Class." *Feminist Studies* 42, no. 1 (2016): 233.

41 Caren Kaplan, "Precision Targets: GPS and the Militarization of U.S. Consumer Identity," *American Quarterly* 58, no. 3 (2006): 707–708.

42 Carl O. Reitz, "UAV / RPV Compendium," UAV Engineering Files, Navy Air Development Center, Warminster, PA, and National Air and Space Museum Archives, Chantilly, VA.

43 Lawrence R. Newcome, *Unmanned Aviation: A Brief History of Unmanned Aerial Vehicles* (Reston, VA: American Institute of Aeronautics and Astronautics, 2004); Michael Armitage, *Unmanned Aircraft* (London: Brassey's Defence Publishers, 1988); Steve Zalonga, *Unmanned Aerial Vehicles: Robotic Air Warfare, 1917–2007*, New Vanguard 144 (Oxford: Osprey Publishing, 2008); Thomas P. Ehrhard, "Air Force UAVs: The Secret History," Mitchell Institute for Airpower Studies, July 2010.

44 Timothy Cullen, "The MQ-9 Reaper Remotely Piloted Aircraft: Humans and Machines in Action" (Ph.D. diss., Massachusetts Institute of Technology, 2011).

45 Michel Foucault, "Nietzsche, Genealogy, History," in *The Foucault Reader*, ed. Paul Rabinow, trans. Donald Bouchard and Sherry Simon (New York: Pantheon Books, 1984), 79.

46 Ibid., 82.

47 "The History of Pilotless Aircraft and Guided Missiles," Collected Records of Delmar Fahrney, Record Group 72 (RG 72), Records of the Bureau of Aeronautics, National Archives at College Park (NARA II), College Park, MD.

48 Ryan Aeronautical Collection, San Diego Air and Space Museum Archives, San Diego, CA.

Chapter 1. DRONE

1 William Standley, from Chief of Naval Operations (CNO) to Naval Bureau of Ordinance (BuOrd), c.c. Bureau of Engineering (BuEng) and Bureau of Aeronautics (BuAer), March 23, 1936, Collected Records of D. S. Fahrney, Records of the Bureau of Aeronautics, RG 72, National Archives at College Park (NARA II), College Park, MD.

2 Fahrney, "History," 201.

3 Joseph Masco, *The Theater of Operations: National Security Affect from the Cold War to the War on Terror* (Durham, NC: Duke University Press, 2014), 3.

4 A. D. Bernhard, from Plans to Chief of BuAer via Asst. Chief of the BuAer, August 19, 1935, Collected Records of D. S. Fahrney, RG 72, NARA II, my emphasis.

5 Samuel Weber, *Targets of Opportunity: On the Militarization of Thinking*, 1st ed. (New York: Fordham University Press, 2005), 4.

6 Williamson Murray wrote that after World War I, "whatever the initial attempts at strategic bombing, both the extent of such attacks and their results left room for considerable debate as to its potential effects on future warfare" ("Strategic Bombing: The British, American, and German Experiences," in *Military Innovation in the Interwar Period*, ed. Allen Millett and Williamson Murray [Cambridge: Cambridge University Press, 1996], 97).

7 Robert Ehlers, *Targeting the Third Reich: Air Intelligence and the Allied Bombing Campaign* (Lawrence: University of Kansas Press, 2009), 17.

8 "Bombs Fail to Sink the *Ostfriesland*," *New York Times*, July 21, 1921, https://www.nytimes.com/1921/07/21/archives/bombs-fail-to-sink-the-ostfriesland-only-four-explode-on-exgerman.html.

9 "Sinking the *Ostfriesland*," *New York Times*, July 23, 1921, 5, https://www.nytimes.com/1921/07/23/archives/sinking-the-ostfriesland.html.

10 Carl Schmitt, *The Nomos of the Earth: In the International Law of Jus Publicum Europaeum*, trans. G. L. Ulmen (New York: Telos Press Publishing, 2006), 319.

11 Ibid., 321.

12 Priya Satia, "Drones: A History from the British Middle East," *Humanity* 5, no. 1 (Spring 2014): 1–31.

13 Clayton Laurie, "The United States Army and the Return to Normalcy in Labor Dispute Interventions: The Case of the West Virginia Coal Mine Wars, 1920–1921," *West Virginia History* 50 (1991): 1–24.

14 Ibid.

15 Ibid.

16 Tami Davis Biddle, *Rhetoric and Reality in Air Warfare: The Evolution of British and American Ideas about Strategic Bombing, 1914–1945*, Princeton Studies in International History and Politics (Princeton, NJ: Princeton University Press, 2002), 6–11.

17 Murray, "Strategic Bombing," 107.

18 See Hugh G. J. Aitken, *The Continuous Wave: Technology and American Radio, 1900–1932* (Princeton, NJ: Princeton University Press, 2016); Susan Jeanne Douglas, *Inventing American Broadcasting: 1899–1922*, Johns Hopkins Studies in the History of Technology (Baltimore: Johns Hopkins University Press, 1989); Sunggok Hong, *Wireless: From Marconi's Black-Box to the Audion*, Transformations: Studies in the History of Science and Technology (Cambridge, MA: MIT Press, 2001).

19 Jussi Parikka, *Insect Media: An Archaeology of Animals and Technology* (Minneapolis: University of Minnesota Press, 2010), xiii.

20 Lucy Suchman, *Human-Machine Reconfigurations: Plans and Situated Actions* (Cambridge: Cambridge University Press, 2007), 12, 23.

21 A. B. Cook, from Chief of the BuAer to the CNO, April 15, 1936, Collected Records of D. S. Fahrney, RG 72, NARA II.

22 Ibid.

23 Bruno Latour, "Where Are the Missing Masses? The Sociology of a Few Mundane Artifacts," in *Shaping Technology, Building Society: Studies in Sociotechnical Change,* ed. Wiebe Bijker and John Law (Cambridge, MA: MIT Press, 1992), 254.

24 Delmar Fahrney, "Monthly Report—September 1936," Naval Research Laboratory, Washington, DC, Collected Records of Delmar Fahrney, RG 72, NARA II.

25 Bruno Latour, *Pandora's Hope: Essays on the Reality of Science Studies*, trans. Catherine Porter (Cambridge, MA: Harvard University Press, 1999), 309.

26 Delmar Fahrney, "Monthly Report—April 1937," Naval Research Laboratory, Washington, DC, Collected Records of Delmar Fahrney, RG 72, NARA II.

27 Ibid.

28 Ibid.

29 Delmar Fahrney, "Monthly Report—November 1937," Naval Research Laboratory, Washington, DC, Collected Records of Delmar Fahrney, RG 72, NARA II.

30 Ibid.

31 Ibid.

32 Ibid.

33 Ibid., my emphasis.

34 Ibid.

35 Fahrney, "History," 213.

36 "La culture se conduit envers l'objet technique comme l'homme envers l'étranger quand il se laisse emporter par la xénophobie primitive." Gilbert Simondon, *Du Mode d'Existence des Objets Techniques* (Paris: Aubier, 1958), 10.

37 "La machine est l'étrangère; c'est l'étrangère en laquelle est enfermé de l'humain, méconnu, matérialisé, asservi, mais restant pourtant de l'humain." Ibid., 10.

38 Fahrney, "History," 213.

39 Qtd. in Fahrney, "History," 230.

40 Delmar Fahrney, letter to Albert G. Noble, August 29, 1938, Collected Records of Delmar Fahrney, RG 72, NARA II.

41 Fahrney, "History," 211–220.

42 Delmar Fahrney, letter to BuAer and Naval Division of Fleet Training, August 13, 1938, Collected Records of Delmar Fahrney, RG 72, NARA II.

43 Ibid.

44 John S. McCain, letter to US Fleet Commander-in-Chief and Battleforce Commander, August 29, 1938, Collected Records of Delmar Fahrney, RG 72, NARA II.

45 Ibid.

46 Ibid.

47 Ibid., my emphasis.

48 Fahrney to Noble, August 29, 1938.

49 McCain to US Fleet Commander-in-Chief and Battleforce Commander, August 29, 1938.

50 Qtd. in Fahrney, "History," 238, my emphasis.

51 Delmar Fahrney, letter to Chief of BuAer, "Procedure for a Dive Bombing Practice by a Drone on the U.S.S. *Utah*," September 11, 1938, Collected Records of Delmar Fahrney, RG 72, NARA II.

52 Walter Brown, letter to Delmar S. Fahrney, US Fleet Commander-in-Chief, and Commander Base Force, September 17, 1938, Collected Records of Delmar Fahrney, RG 72, NARA II.

53 Fahrney, "History," 249.

54 Ibid., 265, my emphasis.

55 Qtd. in ibid., 293.

56 Qtd. in ibid., 295–296.

57 Qtd. in ibid., 296.

2. American Kamikaze

1 Vannevar Bush, "Guided Missile Review" (Washington, DC: Office of Scientific Research and Development, 1947), in Collected Records of Delmar Fahrney, RG 72, National Archives at College Park (NARA II), College Park, MD.

2 Henry Garrett, "Letter from the Secretary of the Navy | Special Task Air Group One (STAG-1)," July 5, 1990, http://stagone.org/?page_id=50.

3 James Hall, *American Kamikaze* (Titusville, FL: J. Bryant, 1984), 54. I also analyze this case in Katherine Chandler, "American Kamikaze: Television-Guided Assault Drones in World War II." In *Life in the Age of Drone Warfare*, edited by Lisa Parks and Caren Kaplan, 89-111. Durham, NC: Duke University Press, 2017.

4   Judith Butler, *Frames of War: When Is Life Grievable?* (London: Verso Books, 2009), 1.

5   Ibid., 4.

6   Ibid., 7, 8.

7   Ibid., 12.

8   Hall, *American Kamikaze*, 55.

9   William Uricchio, "Television's First Seventy-five Years: The Interpretative Flexibility of a Medium in Transition," in *The Oxford Handbook of Film and Media Studies*, ed. Robert Kolker (Oxford: Oxford University Press, 2008), 289.

10 Vladimir Zworykin, "Flying Torpedo with an Electric Eye," 1934, 1, RCA Collection, Hagley Library Manuscripts Collection, Wilmington, DE.

11 Ibid.

12 Ibid.

13 See the end of this chapter, as well as Emiko Ohnuky-Tierney, *Kamikaze, Cherry Blossoms, and Nationalisms: The Militarization of Aesthetics in Japanese History* (Chicago: University of Chicago Press, 2002) for analysis of Japanese U.S. racism reflected in these discussions.

14 Zworykin, "Flying Torpedo," 1–2; my emphasis.

15 Ibid., 2.

16 In the last chapter, I will suggest how this network of parts removed the human in a fashion similar to that described in Paul Virilio's classic formulation of airplane, gun, and camera (*War and Cinema: The Logistics of Perception*, trans. Patrick Camiller, Radical Thinkers [1989; repr., London: Verso Books, 2009]).

17 Rey Chow, *The Age of the World Target: Self-Referentiality in War, Theory and Comparative Work* (Durham, NC: Duke University Press, 2006), 31.

18 Ibid., 41.

19 Qtd. in Fahrney, "History," 318.

20 Claude Bloch, "Commentary from Commander in Chief of the Navy to Chief of the BuAer," Assault Drones (Washington, DC: Navy Chief Naval Officer, 1939), Collected Records of Delmar Fahrney, RG 72, NARA II.

21 Donald A. MacKenzie, *Inventing Accuracy: A Historical Sociology of Nuclear Missile Guidance* (Cambridge, MA: MIT Press, 1993), 214.

22 Ibid., 384.

23 Fahrney, "History."

24 Gary Edgerton, *Columbia History of Television* (New York: Columbia University Press, 2010), 70–71.

25 Walter Webster, "Report from Manager of NAF to Chief of BuAer," Assault Drones (Philadelphia: Naval Aircraft Factory, August 22, 1941), Collected Records of Delmar Fahrney, RG 72, NARA II. In the report, the term "drone" is capitalized (DRONE) to indicate that it is a code name.

26 Fahrney, "History," 338. He explains further, "Since the established aircraft industry could not be used, design called for a plastic plywood airplane powered by the flat air-cooled 150 h.p. engine."

27 Qtd. in Fahrney, "History," 339.

28 For example, see Albert Axell and Hideaki Kase, *Kamikaze: Japan's Suicide Gods* (London: Longman, 2002), 40–44.

29 See Robert G. Lee, *Orientals: Asian Americans in Popular Culture*, Asian American History and Culture (Philadelphia: Temple University Press, 1999); John Kuo Wei Tchen and Dylan Yeats, eds., *Yellow Peril! An Archive of Anti-Asian Fear* (London:

Verso Books, 2014); Erika Lee, *The Making of Asian America: A History* (New York: Simon & Schuster, 2015).

30 Qtd., in Fahrney, "History," 372.

31 Fahrney, "History," 373.

32 "TDR-1 EDNA III," *National Naval Aviation Museum* (blog), n.d., http://www .navalaviationmuseum.org/attractions/aircraft-exhibits/item/?item=tdr.

33 Fahrney, "History," 394.

34 Ibid., 388–396.

35 Ibid., 401.

36 Ibid., 424.

37 Ibid., 404.

38 The motion picture archivist explained to me that the tape was given to the assistant director of the Smithsonian, a former member of the U.S. military, anonymously and that he donated it to the National Air and Space Museum archives rather than the Smithsonian archives. Part of the film is available at *Service Test in Field of TDR1*, www.youtube.com/watch?v=8RQcUtzAe98.

39 On Nolo, see Fahrney, "History," 211–214.

40 Chamayou, *Theory of the Drone*, 84.

41 Fahrney, "History," 416.

42 Army reports refer to the drone as a robot.

43 E. E. Partridge, "Summary of Navy Drone Project," memorandum Major General USA to Commanding General, Eighth Air Force, Report on Aphrodite Project, January 20, 1945, War Weary Willies, Miscellaneous Decimal Files, RG 236, NARA II.

44 Ibid.

45 Ibid.

46 Qtd. in Fahrney, "History," 419.

47 Qtd. in Nick T. Spark, "Command Break: The Battle Over America's Secret WWII Cruise Missile," *Proceedings* (United States Naval Institute) (2005), http://stagone .org/?page_id=20.

48 Hall, *American Kamikaze*, 214.

49 Qtd. in ibid., 214.

50 Ibid.

51 Qtd. in Fahrney, "History," 421.

52 Fahrney, "History," 427.

53 Oscar Smith, Letter to Robert Jones, November 26, 1944, Collected Records of Delmar Fahrney, RG 72, NARA II.

54 Bush, "Guided Missile Review."

55 Delmar Fahrney to P. D. Stroop, November 28, 1960, Delmar Fahrney Technical File, National Air and Space Museum Archive, Chantilly, VA.

56 P. D. Stroop to Delmar Fahrney, January 10, 1961, Delmar Fahrney Technical File, National Air and Space Museum Archive, Chantilly, VA.

57 Lee Pearson, "Interview with Delmar Fahrney," March 17, 1960, Delmar Fahrney Technical File, National Air and Space Museum Archive, Chantilly, VA.

58 See Kenneth Werrell, *The Evolution of the Cruise Missile* (Maxwell Air Force Base, AL: Air University Press, 1985); Newcome, *Unmanned Aviation*.

59 Delmar S. Fahrney, "The Birth of the Guided Missile," *United States Naval Institute Proceedings* (December 1980).

60 Delmar S. Fahrney, "The Genesis of the Cruise Missile," *Astronautics and Aeronautics* 20, no. 1 (1982): 34.

3. Unmanning

1   Qtd. in William Wagner, *Lightning Bugs and Other Reconnaissance Drones: The Can-Do Story of Ryan's Unmanned Spy Plane* (Fallbrook, CA: Aero Publishers; Armed Forces Journal International, 1982), 19.

2   "'Brain Box' of the Air Forces," *Washington Daily News*, February 12, 1945, Records of the Bureau of Aeronautics, Collected Records of Delmar Fahrney, Record Group 72, National Archives at College Park, College Park, MD.

3   Qtd. in Wagner, *Lightning Bugs and Other Reconnaissance Drones*, 20.

4   Paul Edwards, *The Closed World: Computers and the Politics of Discourse in Cold War America* (Cambridge, MA: MIT Press, 1996), 7.

5   See Edwards's account of Operation Igloo White in ibid., 3–8.

6   Masco, *Theater of Operations*, 18.

7   Hannah Arendt, *The Human Condition*, 2nd ed. (Chicago: University of Chicago Press, 1998), 152.

8   Ibid., 151.

9   "The Bee with an Electronic Brain," *Ryan Reporter*, March 15, 1953, 13, A/BQM-34 Technical Files, National Air and Space Museum Archives, Chantilly, VA. The technical files at the National Air and Space Museum (NASM) from which this article is drawn are collections of news clippings, press releases, and photographs. Here, I focus on materials that circulated publicly.

10 Geoffrey Bowker, "How to Be Universal: Some Cybernetic Strategies, 1943–70," *Social Studies of Science* 23, no. 1 (1993): 107–127. Cybernetics is what Bowker calls a "universal science," even as his article points to the ways cybernetics emerges in the immediate aftermath of World War II. Other discussions of cybernetics in this period include: Jean-Pierre Dupuy, *The Mechanization of the Mind: The Origins of Cognitive Science*, trans. M. B. DeBevoise (Princeton, NJ: Princeton University Press, 2000); John Johnston, *The Allure of Machinic Life: Cybernetics, Artificial Life, and the New AI* (Cambridge, MA: MIT Press, 2008).

11 Arturo Rosenblueth, Norbert Wiener, and Julian Bigelow, "Behavior, Purpose, and Teleology," *Philosophy of Science* 10, no. 1 (1943): 19.

12 "The Bee with an Electronic Brain," 12–13.

13 See Hayles, *How We Became Posthuman*; Ronald Kline, " "Where Are the Cyborgs in Cybernetics?" *Social Studies of Science* 39, no. 3 (June 2009): 331–362.

14 "The Bee with an Electronic Brain," 13.

15 Ibid.

16 C. T. Turner and G. R. Cota, "Firebee I—A Case Study in Pilotless Aircraft Evolution," 1981, San Diego Air and Space Museum Archives, San Diego, CA.

17 Rosenblueth, Wiener, and Bigelow, "Behavior, Purpose, and Teleology," 26.

18 "The Bee with an Electronic Brain," 13.

19 "Black Box," *OED Online* (Oxford: Oxford University Press, December 2013), http://www.oed.com/view/Entry/282116?redirectedFrom=black box.

20 MacKenzie, *Inventing Accuracy*, 26.

21 Ibid., 381.

22 "A/BQM-34 Technical Files," Technical Reference Files, National Air and Space Museum Archives, Chantilly, VA.

23 Peter Galison, "The Ontology of the Enemy: Norbert Weiner and the Cybernetic Vision," *Critical Inquiry* 21, no. 2 (2004): 231.

24 Ibid., 265.

25 Ibid. 230.

26 See Denis E. Cosgrove and William L. Fox, *Photography and Flight* (London: Reak-tion Books, 2010); Jeanne Haffner, *The View from Above: The Science of Social Space* (Cambridge, MA: MIT Press, 2013).

27 Richard S. Leghorn, "Aerial Reconnaissance," in *Selected Readings in Aerial Recon-naissance*, ed. Amrom Katz (Santa Monica, CA: Rand Corporation, 1951), 9.

28 Qtd. in Philip Taubman, *Secret Empire: Eisenhower, the CIA, and the Hidden Story of America's Space Espionage* (New York: Simon & Schuster, 2003), 85.

29 James R. Killian, "Report by the Technological Capabilities Panel of the Science Advisory Committee," *Foreign Relations of the United States, 1955–1957* (Wash-ington, DC: Science Advisory Committee, February 14, 1955), *National Security Policy*, Vol. 19, Document 9, Department of State Office of the Historian, https://history.state.gov/historicaldocuments/frus1955-57v19/d9. See also Edward Keefer, Charles Sampson, and Louis Smith, *Cuban Missile Crisis and Aftermath: Foreign Relations of the United States, 1961–1963*, Vol. 9 (Washington, DC: Government Printing Office, 1996).

30 Killian, "Report by the Technological Capabilities Panel of the Science Advisory Committee."

31 Ibid.

32 Kristie Macrakis, "Technophilic Hubris and Espionage Styles during the Cold War," *Isis* 101, no. 2 (June 2010): 378, 383, 381.

33 Taubman, *Secret Empire*, 100–109.

34 Ibid., 103–108.

35 Edwards, *The Closed World*, 8. Other histories that examine the Cold War through a technological lens are, e.g., Thomas P. Hughes, *Rescuing Prometheus* (New York: Pantheon, 1998); and Walter McDougall, *The Heavens and the Earth: A Politi-cal History of the Space Age* (New York: Basic Books, 1984). I rely on Edwards's formulation because the questions raised in his analysis of computers offer a useful counterpoint for my study of drones.

36 Taubman, *Secret Empire*, 187.

37 Ibid., 298–299.

38 Lloyd Ryan, Interview by William Wagner, February 15, 1971, Ryan Aeronautical Files, San Diego Air and Space Museum Archives.

39 Ibid.

40 Robert Schwanhausser later came out as a trans woman, Bobbi Swan. In some interviews after coming out, she continued to use her previous name and he/him pronouns when discussing this period in her life, which I follow in my discussion of her involvement. See Peter Rowe, "His and Hers: Robert Schwanhausser's Life Has Two Big Chapters: One as a Man and Now One as a Woman," *San Diego Union Tribune*, June 17, 2007, http://legacy.sandiegouniontribune.com/uniontrib/20070617/news_mz1c17gender.html; Bobbi Swan, "Whatever Happened to Bobbi Swan?" *TG Forum* (blog), May 23, 2009, https://tgforum.com/wordpress/whatever-happened-to-bobbi-swan; Jacob Bernstein, "For Some in Transgender Community, It's Never Too Late to Make a Change," *New York Times*, March 6, 2015, https://www.nytimes.com/2015/03/08/fashion/for-some-in-transgender-community-its-never-too-late-to-make-a-change.html.

41 Qtd. in Wagner, *Lightning Bugs and Other Reconnaissance Drones*, 13. Note also that Schwanhausser frames this new use for drones as a "logical evolution."

42 Nikita Khrushchev, "Summit Conference Statement," *Department of State Bulletin*, June 6, 1960.

43 Robert Schwanhausser, Interview by William Wagner, n.d., Ryan Aeronautical Files, RRS#1, San Diego Air and Space Museum Archives.

44 Qtd. in Wagner, *Lightning Bugs and Other Reconnaissance Drones*, 18.

45 Wagner, *Lightning Bugs and Other Reconnaissance Drones*, 23.

46 Turner and Cota, "Firebee I," 5.

47 "Reconnaissance: Cameras Aloft: No Secrets Below," *Time*, December 28, 1962, http://content.time.com/time/subscriber/article/0,33009,827960-1,00.html.

48 Numerous published accounts explore the details of the Cuban Missile Crisis, the intelligence that was collected, and the decisions that it precipitated. I point to the event, however, to examine how reconnaissance imagery, which had been considered highly secret in the 1950s, circulated to the U.S. public in this event and the role images play in the Cold War crisis. For further analysis of intelligence operations, see Dino Brugioni, *Eyeball to Eyeball: The Inside Story of the Cuban Missile Crisis* (New York: Random House, 1991); and James Light and David Welch, *Intelligence and the Cuban Missile Crisis* (London: Frank Cass, 1998). The event is also examined in Keefer, Sampson, and Smith, *Cuban Missile Crisis and Aftermath*.

49 John F. Kennedy, "Address on the Cuban Crisis October 22, 1962," *Internet Modern History Sourcebook*, October 22, 1962, https://sourcebooks.fordham.edu/mod/1962kennedy-cuba.asp.

50 Ibid.

51 Macrakis, "Technophilic Hubris and Espionage Styles during the Cold War," 380.

52 Schwanhausser, Interview.

53 Qtd. in Wagner, *Lightning Bugs and Other Reconnaissance Drones*, 50.

54 "History of TRA Government Contracts."

55 "Reconnaissance: Cameras Aloft: No Secrets Below."

56 "Cuban Missile Crisis Briefing by the Defense Intelligence Agency," February 6, 1963, https://www.youtube.com/watch?v=eo6QVTTOf1w.

57 Tom Wicker, "M'Namara Insists Offensive Arms Are Out of Cuba," *New York Times*, February 7, 1963, https://www.nytimes.com/1963/02/07/archives/mnamara-insists-offensive-arms-are-out-of-cuba-declares-on-tv-that.html.

58 Qtd. in ibid.

59 Ibid.

4. Buffalo Hunter

1 Hanson W. Baldwin, "The 'Drone': Portent of Push-Button War; Recent Operations Point to the Pilotless Plane as a Formidable Weapon in War's New Armory," *New York Times*, August 25, 1946, https://www.nytimes.com/1946/08/25/archives/the-drone-portent-of-pushbutton-war-recent-operations-point-to-the.html.

2 *Operation Crossroads Scrapbook* (United States Army Air Forces, 1946), MRC 322, National Air and Space Museum Archives, Chantilly, VA, 9

3 Ibid., 2.

4 Ruth Oldenziel, "Islands: The United States as Networked Empire," in *Entangled Geographies: Empire and Technopolitics in the Global Cold War*, ed. G. Hecht (Cambridge, MA: MIT Press, 2011), 21.

5 Mbembe, "Necropolitics," 24, 25, 29.

6   This section examines images that circulated publicly in 1953. The materials were received by the Institute of the Aeronautical Sciences and were subsequently archived by NASM.

7   Ryan Aeronautical, Photograph, May 21, 1953, A/BQM-34 Technical Files, National Air and Space Museum Archives, Chantilly, VA.

8   "Safety pilot" was the name given to the human pilots who tested the drone aircraft when the systems were developed in the interwar period. See the discussion in chapter 1 for more detail.

9   The military designation Q-2 was later changed to A/BQM-34, which applied to both target drones and the reconnaissance systems modeled on the Firebee. The reconnaissance aircraft were coded through various insect names including Lightning Bug and Firefly. Other model names included Compass Arrow and Compass Cope. In this chapter, I use the name Firebee to refer to both target and reconnaissance drones (much like the military designation).

10 See, for example, Valerie Kuletz, *Tainted Desert: Environmental and Social Ruin in the American West* (New York: Routledge, 1998); Rebecca Solnit, *Savage Dreams: A Journey into the Landscape Wars of the American West* (Berkeley: University of California Press, 2014); Shiloh R. Krupar, *Hot Spotter's Report: Military Fables of Toxic Waste* (Minneapolis: University of Minnesota Press, 2013).

11 A classic account of the Trinity Test is Lansing Lamont, *The Day of Trinity* (New York: Atheneum, 1965); see also Ferenc Szasz, *The Day the Sun Rose Twice: The Story of the Trinity Site Nuclear Explosion, July 16, 1945* (Albuquerque: University of New Mexico Press, 1984).

12 Donna Haraway, "Teddy Bear Patriarchy: Taxidermy in the Garden of Eden, New York City, 1908–1936," *Social Text*, no. 11 (1984): 47.

13 William Grimes, *The History of Big Safari* (New York: Archway Publisher, 2014), 2.

14 Schwanhausser, Interview, 2.

15 "Bird" is a slang term used by Ryan Aeronautical personnel to refer to any aircraft.

16 Schwanhausser, Interview, 2.

17 Ibid., 2, 3.

18 Qtd. in Wagner, *Lightning Bugs*, 30.

19 Qtd. in ibid.

20 Qtd. in ibid.

21 Qtd. in ibid.

22 "Secret Something Falls to Earth," *Albuquerque Journal*, August 6, 1969, Ryan Aeronautical Files, San Diego Air and Space Museum Archives. Cited in William Wagner and William P. Sloan, *Fireflies and Other UAVs (Unmanned Aerial Vehicles)* (Arlington, TX: Aerofax, 1992), 31.

23 Wagner and Sloan, *Fireflies and Other UAVs*, 29. See also "Drone Test Pattern Outlined," *Aerospace Daily*, November 22, 1969; Charles D. La Fond, "Air Force Learns It's Tough to Keep a Secret," *Washington Waveguide*, September 1969; and Bill Stockton, " 'Firefly' Is Shrouded in Secrecy," *Alamagordo Daily News*, August 5, 1969, all in Ryan Aeronautical Files, San Diego Air and Space Museum Archives.

24 Qtd. in Wagner and Sloan, *Fireflies and Other UAVs*, 33.

25 Ibid.

26 Joseph Masco, *Nuclear Borderlands: The Manhattan Project in Post–Cold War New Mexico* (Princeton, NJ: Princeton University Press, 2006), 284.

27 Caren Kaplan, *Aerial Aftermaths: Wartime from Above*, Next Wave (Durham, NC: Duke University Press, 2018), 14.

28 For a detailed analysis of August 24, 1962, and the escalations that resulted because of this incident, see Edwin Moise, *Tonkin Gulf and Escalation of the Vietnam War* (Chapel Hill: University of North Carolina Press, 1996).

29 Wagner and Sloan, *Fireflies and Other UAVs*, 16.

30 Qtd. in Wagner, *Lightning Bugs*, 55.

31 Alex Roland, *The Military-Industrial Complex* (Washington, DC: American Historical Association, 2001), 15.

32 Schwanhausser, Interview, 1.

33 Paul Elder, "Buffalo Hunter," Department of the Air Force: Director of Operations Analysis CHECO/CORONA HARVEST DIVISION, July 24, 1973, 4, https://www.scribd.com/document/81658966/7-24-1973-BUFFALO-HUNTER-U-1970-1972.

34 Qtd. in Wagner, *Lightning Bugs*, 62.

35 Ibid., 63.

36 Ibid., 93. See also "Table 3.3.-1 History of TRA Government Contracts Related to Remotely Piloted Vehicles 1962-1987,." Ryan Aeronautical Collection, San Diego Air and Space Museum Archives, San Diego, CA.

37 Wagner, *Lightning Bugs*, 55.

38 Schmitt, *Nomos of the Earth*, 354.

39 Elder, "Buffalo Hunter," 32.

40 Ibid., xi.

41 Ibid., xii.

42 Ibid., 17.

43 Wagner, *Lightning Bugs*, 204–205.

44 Elder, "Buffalo Hunter," 2.

45 Wagner, *Lightning Bugs*, 136–137.

5. Pioneer

1 CIA (Central Intelligence Agency), "Remotely Piloted Vehicles in the Third World: A New Military Capability," *CIA FOIA*, iii, https://www.cia.gov/library/reading room/document/cia-rdp87t01127r001000830003-3.

2 Ibid., iv.

3 Ibid.

4 Ibid., 5.

5 Robert E. Harkavy, "Pariah States and Nuclear Proliferation," *International Organization* 35, no. 01 (December 1981): 136. https://doi.org/10.1017/S0020818300004112.

6 "The Dumb Pursuit of a Smart Weapon," *New York Times*, March 28, 1988, https://www.nytimes.com/1988/03/28/opinion/the-dumb-pursuit-of-a-smart-weapon.html.

7 "Corruption," *OED Online* (Oxford: Oxford University Press, 2019), http://www.oed.com.proxy.library.georgetown.edu/view/Entry/42045.

8 Virilio, *War and Cinema,* 1.

9 Ibid., 2.

10 Ibid., 5.

11 "Israel, 12.4I," n.d., Ryan Aeronautical Files, San Diego Air and Space Museum Archives.

12 Robert Schwanhausser, Background on the Israeli Program, 5, Interview by William Wagner, November 24, 1971, Ryan Aeronautical Collection, RRS#1, San Diego Air and Space Museum Archives.

13 Wagner and Sloan, *Fireflies and Other UAVs*, 53.

14 I consulted with archivists about finding the export license in the State Department records; these records, however, have been destroyed as is the protocol for military sales.

15 "TRA Drone / RPV Systems Data (Chart)," n.d., Ryan Aeronautical Collection, San Diego Air and Space Museum Archives. An identical chart can be found in "A/ BQM-34 Technical Files."

16 The original article and translation, "First—Unveiling—RPV's in Service of the IAF" are in the file "Israel, 1241," Ryan Aeronautical Collection, San Diego Air and Space Museum Archives.

17 The sale in 1971 of the Ryan Aeronautical reconnaissance drone also coincided with Israeli investments in the development of unmanned aircraft and Israel's purchase of other drones, including a model built by Northrop.

18 Benjamin Schemmer, "Foreword," in *Lightning Bugs and Other Reconnaissance Drones: The Can-Do Story of Ryan's Unmanned Spy Plane*, by William Wagner (Fallbrook, CA: Aero Publishers; Armed Forces Journal International, 1982), n.p.

19 Ibid., emphasis original.

20 Robert Lindsey, "Al Ellis: Have Drone, Will Travel," *New York Times*, October 9, 1983, https://www.nytimes.com/1983/10/09/weekinreview/al-ellis-have-drone-will -travel.html.

21 Ibid.

22 "First UAV Squadron (1971–2007)," *Israeli Air Force* (blog), http://www.iaf.org.il /4968-33518-en/IAF.aspx.

23 Caspar W. Weinberger, *Fighting for Peace: Seven Critical Years in the Pentagon* (New York: Warner Books, 1990), 149. See also the reaction by Dore Gold, "With Friends Like These," *Jerusalem Post*, March 20, 1992.

24 See Israeli Defense Forces (IDF), Research and Development Unit, "Subject: Drones—Transition from Development to Manufacturing," May 29, 1978; "Subject: Drones—Transition from Development to Manufacturing," August 23, 1979; and "Subject: 'Sorek'/"Zahavan"—Purchase Recommendations," February 27, 1980, all trans. Sara Grayson, all IDF & Defense Establishment Archives, Tel-Aviv, Israel.

25 See Zalonga, *Unmanned Aerial Vehicles*, 21–23. Greg Goebel, "[8.0] International Battlefield UAVs," *Greg Goebel/In the Public Domain*, January 1, 2003, http:// craymond.no-ip.info/awk/twuav8.html. The merger is also described in IDF Archive materials.

26 See "Company Overview of Tadiran Air Conditioners Ltd.," *Bloomberg Business* (blog), June 29, 2018, https://www.bloomberg.com/research/stocks/private /snapshot.asp?privcapId=5093074.

27 Ehrhard, "Air Force UAVs," 39–41. Ehrhard carefully situates the development of UAVs within the institutional and organizational context of the military, highlighting how internal dynamics of the air force challenged development of the systems. This work builds on his exhaustive dissertation on UAVs written in 2000. Yet, within the Mitchell Institute, directed in 2010 by Rebecca Grant whom I quote later in this chapter, his argument is nonetheless situated as prelude to the

UAV revolution that occurs after 9/11, highlighting their eventual "astounding success."

28 See Andreas Parsch, "MQM-105," 2002, *Directory of U.S. Military Rockets and Missiles*, http://www.designation-systems.net/dusrm/m-105.htm; Grover L. Alexander, "AQUILA Remotely Piloted Vehicle System Technology Demonstrator (RPV-STD) Program," U.S. Army Aviation Research and Development Command, 1979, https://archive.org/details/DTIC_ADA068345/page/n2.

29 Office of the Deputy of Chief of Staff, "Subject: Drone Sharing in Drill 252—Summary," December 8, 1980, trans. Sara Grayson, IDF & Defense Establishment Archives, Tel-Aviv, Israel.

30 Rebecca Grant, "The Bekaa Valley War," *Air Force Magazine* 85, no. 6 (2002): 58; my emphasis.

31 Ed Magnuson, "Israel Strikes at the PLO," *Time*, June 21, 1982, 12, http://content .time.com/time/magazine/article/0,9171,925452,00.html.

32 Benjamin S. Lambeth, *The Transformation of American Air Power*, Cornell Studies in Security Affairs (Ithaca, NY: Cornell University Press, 2000), 96.

33 Ralph Sanders, "An Israeli Military Innovation: UAVs," *Joint Force Quarterly* 33 (2002): 114.

34 Ibid., 115.

35 Grant, "The Bekaa Valley War," 60.

36 Ibid., 61.

37 "First UAV Squadron (1971–2007)"; see also John Kreis, "Unmanned Aircraft in Israeli Air Operations," *Air Power History* 37, no. 4 (1990): 46–50.

38 "First UAV Squadron (1971–2007)."

39 Matthew Hurley, "The Bekaa Valley Air Battle, June 1982: Lessons Mislearned?" *Airpower Journal* 3, no. 4 (1989): 60–70.

40 "First UAV Squadron (1971–2007)."

41 Magnuson, "Israel Strikes at PLO," 5.

42 Southern Command, "Subject: Drone Operation in the Field Forces—Operation Shalom Hagalil—First Report," July 2, 1982, trans. Sara Grayson, IDF & Defense Establishment Archives, Tel-Aviv, Israel.

43 Ibid.

44 Ibid.

45 Ibid.

46 Office of the Deputy of Chief of Staff, "Subject: Outline of the Drones—Discussion Summary from the Date December 5, 1982," December 14, 1982, trans. Sara Grayson, IDF & Defense Establishment Archives, Tel-Aviv, Israel.

47 Head of the Purchasing and Manufacturing Administration, "Subject: Emergency Purchase of Drones," June 28, 1982, trans. Sara Grayson, IDF & Defense Establishment Archives, Tel-Aviv, Israel.

48 Grant, "The Bekaa Valley War," 58.

49 "The Lebanon War: Operation Peace for Galilee (1982)," *Israeli Missions Around the World* (blog), n.d., http://embassies.gov.il/MFA/AboutIsrael/history/Pages /Operation%20Peace%20for%20Galilee%20-%201982.aspx.

50 United Nations General Assembly, "The Situation in the Middle East," December 16, 1982, https://www.un.org/documents/ga/res/37/a37r123.htm.

51 Weinberger, *Fighting for Peace*, 148–149.

52 Gold, "With Friends Like These."

53 Akhil Gupta, "Blurred Boundaries: The Discourse of Corruption, the Culture of Politics, and the Imagined State," *American Ethnologist* 22, no. 2 (May 1995): 387, https://doi.org/10.1525/ae.1995.22.2.02a00090.

54 "The Dumb Pursuit of a Smart Weapon," *New York Times*, March 28, 1988, https://www.nytimes.com/1988/03/28/opinion/the-dumb-pursuit-of-a-smart-weapon.html.

55 Ibid.

56 All quotations from Joe Pichirallo, " 'Extraordinary' Process Used to Buy Navy Drone," *Washington Post*, July 28, 1988.

57 Associated Press, "Consultant Admits Bribe of Navy Brass: Defense: Defendant in Justice Department Probe Fingers U.S. Officials and Israeli Businessmen in Contract Conspiracy," *Los Angeles Times*, May 28, 1990.

58 Louis J. Rodrigues, "Unmanned Aerial Vehicles: DOD's Acquisition Efforts," Testimony before Committee on National Security, U.S. House of Representatives (Washington, DC: Government Accountability Office, April 9, 1997), http://careersdocbox.com/US_Military/73538263-Gao-vehicles-unmanned-aerial-dod-s-acquisition-efforts.html.

59 Ibid.

60 Clyde Haberman, "Israelis Kill Chief of Pro-Iran Shiites in South Lebanon," *New York Times*, February 17, 1992, https://www.nytimes.com/1992/02/17/world/israelis-kill-chief-of-pro-iran-shiites-in-south-lebanon.html.

61 "40th Anniversary of the 200th Squadron," trans. Dafna Ruppin (Herzliya, Israel: Israel Air Force Association, n.d.), Fisher Institute for Air and Space Strategic Studies Archives, Herzliya, Israel.

62 Ibid., 19.

63 Qtd. in Haberman, "Israelis Kill Chief of Pro-Iran Shiites in South Lebanon."

64 Qtd. in ibid.

65 Qtd. in ibid.

66 Qtd. in ibid.

67 See Ronen Bergman and Ronnie Hope, *Rise and Kill First: The Secret History of Israel's Targeted Assassinations* (New York: Random House, 2018).

68 Lisa Hajjar, "Lawfare and Armed Conflict: Comparing Israeli and US Targeted Killing Policies and Challenges against Them," Research Report (Beirut: American University of Beirut–Issam Fares Institute for Public Policy and International Affairs, January 2013), http://website.aub.edu.lb/ifi/international_affairs/documents/20130129ifi_pc_ia_research_report_lawfare.pdf.65.

69 *MND-B Soldiers Kill Two Terrorists* (Defense Video and Imagery Distribution System, 2008), https://www.dvidshub.net/video/37449/mnd-b-soldiers-kill-two-terrorists#.U2-TGnavj54.

Conclusion. Nobody's Perfect

1 Wagner and Sloan, *Fireflies and Other UAVs*, 53.

2 Donna Haraway, "A Cyborg Manifesto," in *Simians, Cyborgs, and Natures: The Reinvention of Nature* (New York: Routledge, 1991), 149.

3 Chris Woods and Christina Lamb, "CIA Tactics in Pakistan Include Targeting Rescuers and Funerals," *Bureau of Investigative Journalism*, February 4, 2012, https://www.thebureauinvestigates.com/stories/2012-02-04/cia-tactics-in-pakistan-include-targeting-rescuers-and-funerals.

4 Robert T. Holden, "The Contagiousness of Aircraft Hijacking," *American Journal of Sociology* 91, no. 4 (January 1986): 881, https://doi.org/10.1086/228353.

5 Hugh Gusterson, *Drone: Remote Control Warfare* (Cambridge, MA: MIT Press, 2016), 3.

6 Emily Chow, Alberto Cuadra, and Craig Whitlock, "Fallen from the Skies," *Washington Post*, January 19, 2016, https://www.washingtonpost.com/graphics/national/drone-crashes/database/??noredirect=on.

7 Valerie Insinna, "Number of Air Force Drone-Related Mishaps Has Remained Steady since 2011," *Military Times*, May 15, 2018, https://www.militarytimes.com/news/your-military/aviation-in-crisis/2018/05/15/number-of-air-force-drone-related-mishaps-has-remained-steady-since-2011/.

Bibliography

Archives

Fisher Institute for Air and Space Strategic Studies Archives, Herzliya, Israel

"40th Anniversary of the 200th Squadron." Translated by Dafna Ruppin. Herzliya, Israel: Israel Air Force Association, n.d.

Hagley Library Manuscripts Collection, Wilmington, DE

Zworykin, Vladimir. "Flying Torpedo with an Electric Eye," 1934. RCA Collection.

The IDF & Defense Establishment Archives, Tel-Aviv, Israel

Air Force Commander. "Subject: Summary Discussion of the Operational Concept for Running the Zahavan," April 5, 1978. Translated by Sara Grayson.
Head of the Purchasing and Manufacturing Administration. "Subject: Emergency Purchase of Drones," June 28, 1982. Translated by Sara Grayson.
Office of the Deputy of Chief of Staff. "Subject: Drone Sharing in Drill 252—Summary," December 8, 1980. Translated by Sara Grayson.
———. "Subject: Outline of the Drones—Discussion Summary from the Date December 5, 1982," December 14, 1982. Translated by Sara Grayson.
Research and Development Unit. "Subject: Drones—Transition from Development to Manufacturing," May 29, 1978. Translated by Sara Grayson.
———. "Subject: Drones—Transition from Development to Manufacturing," August 23, 1979. Translated by Sara Grayson.
———. "Subject: 'Sorek'/"Zahavan"—Purchase Recommendations," February 27, 1980. Translated by Sara Grayson.
Southern Command. "Subject: Drone Operation in the Field Forces—Operation Shalom Hagalil—First Report," July 2, 1982. Translated by Sara Grayson.

National Air and Space Museum Archives, Chantilly, VA

"A/BQM-34 Technical Files." Technical Reference Files.
"The Bee with an Electronic Brain." *Ryan Reporter*, March 15, 1953. A/BQM-34 Technical Files.
Fahrney, Delmar. Letter to P. D. Stroop, November 27, 1960. Delmar Fahrney Technical File.

" 'The Father of the Guided Missile' Retires." U.S. Naval Air Missile Test Center, October 30, 1950. Technical Reference Files.

Operation Crossroads Scrapbook. United States Army Air Forces, 1946. MRC 322.

Pearson, Lee. "Interview with Delmar Fahrney," March 17, 1960. Delmar Fahrney Technical File.

"Radioplane Technical Files." Technical Reference Files.

Reitz, Carl O. "UAV / RPV Compendium." Warminster, PA: Navy Air Development Center, 1988. UAV Engineering Files.

Stroop, P. D. Letter to Delmar Farhney, January 10, 1961. Delmar Fahrney Technical File.

National Archives at College Park (NARA II), College Park, MD

Bernhard, A. D. "Plans to Chief of BuAer via Asst. Chief of the BuAer." August 19, 1935. Records of the Bureau of Aeronautics, Collected Records of Delmar Fahrney, Record Group 72 (RG 72).

Bloch, Claude. "Commentary from Commander-in-Chief of the Navy to Chief of the BuAer." Assault Drones. Washington, DC: Navy Chief Naval Officer, 1939. Records of the Bureau of Aeronautics, Collected Records of Delmar Fahrney, RG 72.

" 'Brain Box' of the Air Forces." *Washington Daily News*, February 12, 1945. Records of the Bureau of Aeronautics, Collected Records of Delmar Fahrney, RG 72.

Brown, Walter. Letter to Delmar S. Fahrney, US Fleet Commander-in-Chief, and Commander Base Force, September 17, 1938. Records of the Bureau of Aeronautics, Collected Records of Delmar Fahrney, RG 72.

Bush, Vannevar. "Guided Missile Review." Washington, DC: Office of Scientific Research and Development, 1947. Records of the Bureau of Aeronautics, Collected Records of Delmar Fahrney, RG 72.

Cook, A. B. "Chief of the BuAer to the CNO." April 15, 1936. Records of the Bureau of Aeronautics, Collected Records of Delmar Fahrney, RG 72.

Denny, Reginald. Letter to Delmar S. Fahrney, January 6, 1958. Records of the Bureau of Aeronautics, Collected Records of Delmar Fahrney, RG 72.

Fahrney, Delmar S. "The History of Pilotless Aircraft and Guided Missiles." Records of the Bureau of Aeronautics, Collected Records of Delmar Fahrney, RG 72.

———. Letter to Chief of BuAer. "Procedure for a Dive Bombing Practice by a Drone on the U.S.S. Utah," September 11, 1938. Records of the Bureau of Aeronautics, Collected Records of Delmar Fahrney, RG 72.

———. Letter to Albert G. Noble, August 29, 1938. Records of the Bureau of Aeronautics, Collected Records of Delmar Fahrney, RG 72.

———. Letter to BuAer and Naval Division of Fleet Training, August 13, 1938. Records of the Bureau of Aeronautics, Collected Records of Delmar Fahrney, RG 72.

———. Letter to Naval Bureau of Aeronautics (BuAer), June 21, 1939. Records of the Bureau of Aeronautics, Collected Records of Delmar Fahrney, RG 72.

———. "Monthly Report—April 1937." Navy Research Laboratories, Washington, DC. Records of the Bureau of Aeronautics, Collected Records of Delmar Fahrney, RG 72.

———. "Monthly Report—November 1937." Navy Research Laboratories, Washington, DC. Records of the Bureau of Aeronautics, Collected Records of Delmar Fahrney, RG 72.

———. "Monthly Report—September 1936." Navy Research Laboratories, Washington, DC. Records of the Bureau of Aeronautics, Collected Records of Delmar Fahrney, RG 72.

King, Ernest. "Approval from Chief of the BuAer to Navy CNO." Assault Drones. Washington, DC: Navy Chief Naval Officer, August 22, 1940. Records of the Bureau of Aeronautics, Collected Records of Delmar Fahrney, RG 72.

McCain, John S. Letter to U.S. Fleet Commander-in-Chief and Battleforce Commander, August 29, 1938. Records of the Bureau of Aeronautics, Collected Records of Delmar Fahrney, RG 72.

Miller, C. L. "Visit to U.S. Army Material Division, Wright Field, Dayton, OH" to Plans, July 20, 1936. Washington, DC. Target Drones. Records of the Bureau of Aeronautics, Collected Records of Delmar Fahrney, RG 72.

Partridge, E. E. "Summary of Navy Drone Project." Memorandum Major General USA to Commanding General, Eighth Air Force, Report on Aphrodite Project, January 20, 1945. War Weary Willies, Miscellaneous Decimal Files, Record Group 236.

Powers, E. M., and R. Fink. "Acceptance Tests of Denny Radio Controlled Target Plane." Air Corps, Material Division, October 18, 1938. Records of the Bureau of Aeronautics, Collected Records of Delmar Fahrney, RG 72.

Smith, Oscar. Letter to Robert Jones, November 26, 1944. Records of the Bureau of Aeronautics, Collected Records of Delmar Fahrney, RG 72.

Standley, William. Letter to BuOrd, BuAer, and BuEng, March 23, 1936. Records of the Bureau of Aeronautics, Collected Records of Delmar Fahrney, RG 72.

Webster, Walter. "Report from Manager of NAF to Chief of BuAer." Assault Drones. Philadelphia: Naval Aircraft Factory, August 22, 1941. Records of the Bureau of Aeronautics, Collected Records of Delmar Fahrney, RG 72.

Papers of Dwight D. Eisenhower as President, 1953–1961

Eisenhower, Dwight D. "Farewell Address by President Dwight D. Eisenhower," January 17, 1961. Box 38, Speech Series. https://www.ourdocuments.gov/doc.php?flash=true&doc=90.

San Diego Air and Space Museum Archives, San Diego, CA

Ryan Aeronautical Collection

"Drone Test Pattern Outlined." *Aerospace Daily*, November 22, 1969.

"First—Unveiling—RPV's in Service of the IAF." *IAF Journal* (1974).

"Israel, 124I," n.d.

La Fond, Charles D. "Air Force Learns It's Tough to Keep a Secret." *Washington Waveguide*, September 1969.

Nobody's Perfect. Ryan Aeronautical, n.d.

Ryan, Lloyd. Interview by William Wagner, February 15, 1971.

Schwanhausser, Robert. Background on the Israeli Program. Interview by William Wagner, November 24, 1971. RRS#1.

———. Interview by William Wagner, n.d. RRS#1.

"Secret Something Falls to Earth." *Albuquerque Journal*, August 6, 1969.

Stockton, Bill. "'Firefly' Is Shrouded in Secrecy." *Alamagordo Daily News*, August 5, 1969.

"Table 3.3.-1 History of TRA Government Contracts Related to Remotely Piloted Vehicles 1962–1987," n.d.

"TRA Drone / RPV Systems Data (Chart)," n.d.

Turner, C. T., and G. R. Cota. "Firebee I—A Case Study in Pilotless Aircraft Evolution," 1981.

Vanderbilt Television News Archive, Nashville, TN

"Afghanistan/Al Qaeda/Missile Strike." *CNN Evening News*, February 6, 2002. Vanderbilt Television News Archive. http://tvnews.vanderbilt.edu/diglib-fulldisplay.pl?SID=20140427532989501&code=tvn&RC=656391&Row=4.

"Afghanistan/War on Terror/Missile Attack." *NBC Evening News*, February 7, 2002. Vanderbilt Television News Archive. http://tvnews.vanderbilt.edu/diglib-fulldisplay.pl?SID= 20140427532989501&code=tvn&RC=654761&Row=9.

"Afghanistan/War on Terror/Predator Attack." *CBS Evening News*, February 6, 2002. Vanderbilt Television News Archive. http://tvnews.vanderbilt.edu/diglib-fulldisplay.pl ?SID=20140427532989501&code=tvn&RC=653235&Row=4.

"America Fights Back/Afghanistan/Bin Laden Hunt/Missile Attack." *CBS Evening News*, February 8, 2002. Vanderbilt Television News Archive. http://tvnews.vanderbilt.edu /diglib-fulldisplay.pl?SID=20140427532989501&code=tvn&RC=653272&Row=13.

Other Sources

Abramson, Albert. *Zworykin—Pioneer of Television*. Champaign: University of Illinois Press, 1995.

Aeronautical Oddities. Periscope Film LLC, 2014. https://www.youtube.com/watch?v= SNtA1cbIUSA.

Airwars. "Civilian and 'Friendly Fire' Casualties." London: Airwars, n.d. https://airwars.org /civilian-casualty-claims/.

Aitken, Hugh G. J. *The Continuous Wave: Technology and American Radio, 1900–1932*. 2nd ed. Princeton, NJ: Princeton University Press, 2016.

Alexander, Grover L. "AQUILA Remotely Piloted Vehicle System Technology Demonstrator (RPV-STD) Program." Sunnyvale, CA: U.S. Army Aviation Research and Development Command, 1979. https://archive.org/details/DTIC_ADA068345/page/n2.

Arendt, Hannah. *The Human Condition*. 2nd ed. Chicago: University of Chicago Press, 1998.

Armitage, Michael. *Unmanned Aircraft*. London: Brassey's Defence Publishers, 1988.

Asad, Talal. *On Suicide Bombing*. New York: Columbia University Press, 2007.

Associated Press. "Consultant Admits Bribe of Navy Brass: Defense: Defendant in Justice Department Probe Fingers U.S. Officials and Israeli Businessmen in Contract Conspiracy." *Los Angeles Times*, May 28, 1990.

Axell, Albert, and Hideaki Kase. *Kamikaze: Japan's Suicide Gods*. London: Longman, 2002.

Baldwin, Hanson W. "The 'Drone': Portent of Push-Button War; Recent Operations Point to the Pilotless Plane as a Formidable Weapon in War's New Armory." *New York Times*, August 25, 1946. https://www.nytimes.com/1946/08/25/archives/the-drone-portent-of -pushbutton-war-recent-operations-point-to-the.html.

Barber, Stuart B. "Naval Aviation Combat Statistics—World War II." Air Branch, Office of Naval Intelligence, Office of the Chief of Naval Operations, June 17, 1946. https:// www.history.navy.mil/content/dam/nhhc/research/histories/naval-aviation/aviation -monographs/nasc.pdf.

Barlow, Edward J. *RAND Air Defense Analysis: (A Description of the Structure and Assumptions of the Analysis)*. Santa Monica, CA: RAND, 1951.

Bayard de Volo, Lorraine. "Unmanned? Gender Recalibrations and the Rise of Drone Warfare." *Politics & Gender* 12, no. 1 (2016): 50–77.

Bergman, Ronen, and Ronnie Hope. *Rise and Kill First: The Secret History of Israel's Targeted Assassinations*. New York: Random House, 2018

Bernstein, Jacob. "For Some in Transgender Community, It's Never Too Late to Make a Change." *New York Times*, March 6, 2015. https://www.nytimes.com/2015/03/08 /fashion/for-some-in-transgender-community-its-never-too-late-to-make-a-change .html.

Biddle, Tami Davis. *Rhetoric and Reality in Air Warfare: The Evolution of British and American Ideas about Strategic Bombing, 1914–1945.* Princeton Studies in International History and Politics. Princeton, NJ: Princeton University Press, 2002.

Bijker, Wiebe, Trevor Pinch, and Thomas Hughes, eds. *The Social Construction of Technological Systems: New Directions in the Sociology and History of Technology.* Cambridge, MA: MIT Press, 1989.

"Bombs Fail to Sink the *Ostfriesland.*" *New York Times,* July 21, 1921. https://www.nytimes .com/1921/07/21/archives/bombs-fail-to-sink-the-ostfriesland-only-four-explode-on -exgerman.html.

Borger, Julian. "US Air Wars under Trump: Increasingly Indiscriminate, Increasingly Opaque." *The Guardian,* January 23, 2018. https://www.theguardian.com/us-news/2018 /jan/23/us-air-wars-trump.

Bowker, Geoffrey. "How to Be Universal: Some Cybernetic Strategies, 1943–70." *Social Studies of Science* 23, no. 1 (1993): 107–127.

Brugioni, Dino. *Eyeball to Eyeball: The Inside Story of the Cuban Missile Crisis.* New York: Random House, 1991.

Burns, John F. "A Nation Challenged: The Manhunt: U.S. Leapt before Looking, Angry Villagers Say." *New York Times,* February 17, 2002. https://www.nytimes.com/2002/02/17/world /a-nation-challenged-the-manhunt-us-leapt-before-looking-angry-villagers-say.html.

Butler, Judith. *Frames of War: When Is Life Grievable?* London: Verso Books, 2009.

Butler, Judith, and Biddy Martin. "Cross-Identifications." *Diacritics* 24, no. 2/3 (1994): 3.

Chamayou, Grégoire. *A Theory of the Drone.* Translated by Janet Lloyd. New York: New Press, 2015.

Chandler, Katherine. "5,000 Feet Is the Best: Re-viewing the Politics of Unmanned Aircraft Systems." *Knowledge Politics and Intercultural Dynamics* 5, no. 1 (2012): 63–74.

———. "A Drone Manifesto: Re-forming the Partial Politics of Targeting Killing." *Catalyst* 2, no. 1 (2016): 1–23.

———. "American Kamikaze: Television-Guided Assault Drones in World War II." In *Life in the Age of Drone Warfare,* edited by Lisa Parks and Caren Kaplan, 89–111. Durham, NC: Duke University Press, 2017.

Chow, Emily, Alberto Cuadra, and Craig Whitlock. "Fallen from the Skies." *Washington Post,* January 19, 2016. https://www.washingtonpost.com/graphics/national/drone -crashes/database/.

Chow, Rey. *The Age of the World Target: Self-Referentiality in War, Theory and Comparative Work.* Durham, NC: Duke University Press, 2006.

CIA (Central Intelligence Agency). "Remotely Piloted Vehicles in the Third World: A New Military Capability." *CIA FOIA.* https://www.cia.gov/library/readingroom/document /cia-rdp87t01127r001000830003-3.

Cloud, David S. "Transcripts of U.S. Drone Attack." *Los Angeles Times,* April 8, 2011. http:// documents.latimes.com/transcript-of-drone-attack/.

Cloud, John. "Imaging the World in a Barrel: CORONA and the Clandestine Convergence of the Earth Sciences." *Social Studies of Science* 31, no. 2 (2001): 231–251.

"Company Overview of Tadiran Air Conditioners Ltd." *Bloomberg Business* (blog), June 29, 2018. https://www.bloomberg.com/research/stocks/private/snapshot.asp?privcapId= 5093074.

Cosgrove, Denis E., and William L. Fox. *Photography and Flight.* Exposures. London: Reaktion Books, 2010.

"Creech Air Force Base—Fact Sheet." *Nellis Air Force Base* (blog), July 12, 2012. http://www .nellis.af.mil/About/Fact-Sheets/Display/Article/284172/creech-air-force-base/.

Cullen, Timothy. "The MQ-9 Reaper Remotely Piloted Aircraft: Humans and Machines in Action." Ph.D. diss., Massachusetts Institute of Technology, 2011. http://18.7.29.232/bitstream/handle/1721.1/80249/836824271.pdf?sequence=1.

Daggett, Cara. "Drone Disorientations: How 'Unmanned' Weapons Queer the Experience of Killing in War." *International Feminist Journal of Politics* 17, no. 3 (2015): 361–379. https://doi.org/10.1080/14616742.2015.1075317.

Dao, James. "Raid's Aftermath." *New York Times*, February 12, 2002. https://www.nytimes.com/2002/02/12/world/nation-challenged-raid-s-aftermath-us-defends-missile-strike-saying-attack-was.html.

Der Derian, James. *Virtuous War: Mapping the Military-Industrial-Media-Entertainment Network*. Boulder, CO: Westview Press, 2001.

Douglas, Susan Jeanne. *Inventing American Broadcasting: 1899–1922*. Johns Hopkins Studies in the History of Technology. Baltimore: Johns Hopkins University Press, 1989.

"The Dumb Pursuit of a Smart Weapon." *New York Times*, March 28, 1988. https://www.nytimes.com/1988/03/28/opinion/the-dumb-pursuit-of-a-smart-weapon.html.

Dupuy, Jean-Pierre. *The Mechanization of the Mind: The Origins of Cognitive Science*. Translated by M. B. DeBevoise. Princeton, NJ: Princeton University Press, 2000.

Edgerton, Gary. *Columbia History of Television*. New York: Columbia University Press, 2010.

Edwards, Paul. *The Closed World: Computers and the Politics of Discourse in Cold War America*. Cambridge, MA: MIT Press, 1996.

Ehlers, Robert. *Targeting the Third Reich: Air Intelligence and the Allied Bombing Campaign*. Lawrence: University of Kansas Press, 2009.

Ehrhard, Thomas P. "Air Force UAVs: The Secret History." Mitchell Institute for Airpower Studies, July 2010. https://apps.dtic.mil/dtic/tr/fulltext/u2/a526045.pdf.

———. "Unmanned Aerial Vehicles in the United States Armed Services: A Comparative Study of Weapon System Innovation." Ph.D. diss., Johns Hopkins University, 2000.

Elder, Paul. "Buffalo Hunter." Department of the Air Force: Director of Operations Analysis CHECO/CORONA HARVEST DIVISION, July 24, 1973. https://www.scribd.com/document/81658966/7-24-1973-BUFFALO-HUNTER-U-1970-1972.

Fahrney, Delmar S. "The Birth of the Guided Missile." *United States Naval Institute Proceedings* (December 1980).

———. "The Genesis of the Cruise Missile." *Astronautics and Aeronautics* 20, no. 1 (1982).

Fanon, Frantz. *The Wretched of the Earth*. Translated by Richard Philcox. New York: Grove Press, 2004.

Feldman, Keith. "Empire's Verticality: The Af/Pak Frontier, Visual Culture, and Racialization from Above." *Comparative American Studies* 9, no. 4 (2011): 325–341.

Finn, Peter. "Rise of the Drone: From Calif. Garage to Multibillion-Dollar Defense Industry." *Washington Post*, December 23, 2011. https://www.washingtonpost.com/national/national-security/rise-of-the-drone-from-calif-garage-to-multibillion-dollar-defense-industry/2011/12/22/gIQACG8UEP_story.html?utm_term=.7949ec823efb.

"First UAV Squadron (1971–2007)." *Israeli Air Force* (blog). http://www.iaf.org.il/4968-33518-en/IAF.aspx.

Foucault, Michel. "Nietzsche, Genealogy, History." In *The Foucault Reader*, edited by Paul Rabinow, translated by Donald Bouchard and Sherry Simon, 76–100. New York: Pantheon Books, 1984.

Freud, Sigmund. "The Infantile Genital Organization of the Libido [1923]." In *Collected Papers*, vol. 2, translated by Joan Riviere. New York: Basic Books, 1959.

Galison, Peter. "The Ontology of the Enemy: Norbert Weiner and the Cybernetic Vision." *Critical Inquiry* 21, no. 2 (2004): 228–266.

Garrett, Henry. "Letter from the Secretary of the Navy | Special Task Air Group One (STAG-1)," July 5, 1990. Accessed February 24, 2019. http://stagone.org/?page_id=50.

Gitelman, Lisa. *Always Already New: Media, History, and the Data of Culture*. Cambridge, MA: MIT Press, 2008.

Goebel, Greg. "[8.0] International Battlefield UAVs." *Greg Goebel/In the Public Domain*, January 1, 2003. http://craymond.no-ip.info/awk/twuav8.html.

Gold, Dore. "With Friends Like These." *Jerusalem Post*, March 20, 1992.

Gordon, Avery. *Ghostly Matters: Haunting and the Sociological Imagination*. New ed. Minneapolis: University of Minnesota Press, 2008.

Grant, Rebecca. "The Bekaa Valley War." *Air Force Magazine* 85, no. 6 (2002): 55–61.

Gregory, Derek. "From a View to a Kill: Drones and Late Modern War." *Theory, Culture & Society* 28, no. 7–8 (2011): 188–215.

———. "The Territory of the Screen." *Media Tropes* 6, no. 2 (2016). https://mediatropes.com/index.php/Mediatropes/article/view/28054.

Griffin, Charles. "New Light on Eisenhower's Farewell Address." *Presidential Studies Quarterly* 22, no. 3 (1992): 469–479.

Grimes, William. *The History of Big Safari*. New York: Archway Publisher, 2014.

Gupta, Akhil. "Blurred Boundaries: The Discourse of Corruption, the Culture of Politics, and the Imagined State." *American Ethnologist* 22, no. 2 (May 1995): 375–402. https://doi.org/10.1525/ae.1995.22.2.02a00090.

Gusterson, Hugh. *Drone: Remote Control Warfare*. Cambridge, MA: MIT Press, 2016.

Haberman, Clyde. "Israelis Kill Chief of Pro-Iran Shiites in South Lebanon." *New York Times*, February 17, 1992. https://www.nytimes.com/1992/02/17/world/israelis-kill-chief-of-pro-iran-shiites-in-south-lebanon.html.

Haffner, Jeanne. *The View from Above: The Science of Social Space*. Cambridge, MA: MIT Press, 2013.

Hajjar, Lisa. "Lawfare and Armed Conflict: Comparing Israeli and US Targeted Killing Policies and Challenges against Them." Research Report. Beirut: American University of Beirut–Issam Fares Institute for Public Policy and International Affairs, January 2013. http://website.aub.edu.lb/ifi/international_affairs/documents/20130129ifi_pc_ia_research_report_lawfare.pdf.

Hall, James. *American Kamikaze*. Titusville, FL: J. Bryant, 1984.

Haraway, Donna. "A Cyborg Manifesto." In *Simians, Cyborgs, and Natures: The Reinvention of Nature*. New York: Routledge, 1991.

———. "Situated Knowledges: The Science Question in Feminism and the Privilege of Partial Perspective." *Feminist Studies* 14, no. 3 (1988): 575–599.

———. "Teddy Bear Patriarchy: Taxidermy in the Garden of Eden, New York City, 1908–1936." *Social Text*, no. 11 (1984): 47.

Harkavy, Robert E. "Pariah States and Nuclear Proliferation." *International Organization* 35, no. 01 (December 1981).

Hayles, Katherine. *How We Became Posthuman: Virtual Bodies in Cybernetics, Literature, and Informatics*. Chicago: University of Chicago Press, 1999.

Holden, Robert T. "The Contagiousness of Aircraft Hijacking." *American Journal of Sociology* 91, no. 4 (January 1986): 874–904.

Hong, Sungook. *Wireless: From Marconi's Black-Box to the Audion*. Transformations: Studies in the History of Science and Technology. Cambridge, MA: MIT Press, 2001.

Hughes, Thomas P. *Rescuing Prometheus*. New York: Pantheon, 1998.

Hurley, Matthew M. "The Bekaa Valley Air Battle, June 1982: Lessons Mislearned?" *Airpower Journal* 3, no. 4 (1989): 60–70.

Insinna, Valerie. "Number of Air Force Drone-Related Mishaps Has Remained Steady since 2011." *Military Times*, May 15, 2018. https://www.militarytimes.com/news/your-military /aviation-in-crisis/2018/05/15/number-of-air-force-drone-related-mishaps-has-remained -steady-since-2011/.

"Into the Wild Blue Electronically." *Time*, June 21, 1982. http://content.time.com/time /magazine/article/0,9171,925454,00.html.

Israel. Ministry of Foreign Affairs. "Kehan Report: Report of the Commission of Inquiry into the Events at the Refugee Camps in Beirut—8 February 1983." http://www.mfa.gov .il/mfa/foreignpolicy/mfadocuments/yearbook6/pages/104%20report%20of%20the %20commission%20of%20inquiry%20into%20the%20e.aspx.

Jasanoff, Sheila, ed. *States of Knowledge: The Co-production of Science and the Social Order*. New York, NY: Routledge, 2004.

Johnston, David, and David E. Sanger. "Threats and Responses: Hunt for Suspects; Fatal Strike in Yemen Was Based on Rules Set Out by Bush." *New York Times*, November 6, 2002. https://www.nytimes.com/2002/11/06/world/threats-responses-hunt-for -suspects-fatal-strike-yemen-was-based-rules-set-bush.html.

Johnston, John. *The Allure of Machinic Life: Cybernetics, Artificial Life, and the New AI*. Cambridge, MA: MIT Press, 2008.

Kaplan, Caren. *Aerial Aftermaths: Wartime from Above*. Next Wave. Durham, NC: Duke University Press, 2018.

———. "Precision Targets: GPS and the Militarization of U.S. Consumer Identity." *American Quarterly* 58, no. 3 (2006): 693–714

Keefer, Edward, Charles Sampson, and Louis Smith. *Cuban Missile Crisis and Aftermath: Foreign Relations of the United States, 1961–1963*. Vol. 11. Washington, DC: Government Printing Office, 1996.

Kennedy, John F. "Address on the Cuban Crisis October 22, 1962." *Internet Modern History Sourcebook*, October 22, 1962. https://sourcebooks.fordham.edu/mod/1962kennedy -cuba.asp.

Khrushchev, Nikita. "Summit Conference Statement." *Department of State Bulletin*, June 6, 1960.

Killian, James R. "Report by the Technological Capabilities Panel of the Science Advisory Committee," February 14, 1955. *Foreign Relations of the United States, 1955–1957: National Security Policy*, Vol. 19, Document 9. Washington, DC: Department of State, Office of the Historian. https://history.state.gov/historicaldocuments/frus1955-57v19/d9.

Kindervater, Katharine Hall. "The Emergence of Lethal Surveillance: Watching and Killing in the History of Drone Technology." *Security Dialogue* 47, no. 3 (June 2016): 223–238.

Kittler, Friedrich A. *Gramophone, Film, Typewriter*. Stanford, CA: Stanford University Press, 1999.

Kline, Ronald. "Where Are the Cyborgs in Cybernetics?" *Social Studies of Science* 39, no. 3 (June 2009): 331–362.

Koistinen, Paul. *The Military-Industrial Complex: A Historical Perspective*. New York: Praeger, 1980.

Kreis, John. "Unmanned Aircraft in Israeli Air Operations." *Air Power History* 37, no. 4 (1990): 46–50.

Krupar, Shiloh R. *Hot Spotter's Report: Military Fables of Toxic Waste*. Minneapolis: University of Minnesota Press, 2013.

Kuletz, Valerie. *The Tainted Desert: Environmental Ruin in the American West*. New York: Routledge, 1998.

Lambeth, Benjamin S. *The Transformation of American Air Power*. Cornell Studies in Security Affairs. Ithaca, NY: Cornell University Press, 2000.

Lamont, Lansing. *Day of Trinity*. New York: Atheneum, 1965.

Latour, Bruno. *Pandora's Hope: Essays on the Reality of Science Studies*. Translated by Catherine Porter. Cambridge, MA: Harvard University Press, 1999.

———. *We Have Never Been Modern*. Translated by Catherine Porter. Cambridge, MA: Harvard University Press, 1993.

———. "Where Are the Missing Masses? The Sociology of a Few Mundane Artifacts." In *Shaping Technology, Building Society: Studies in Sociotechnical Change*, edited by Wiebe Bijker and John Law, 225–259. Cambridge, MA: MIT Press, 1992.

Laurie, Clayton. "The United States Army and the Return to Normalcy in Labor Dispute Interventions: The Case of the West Virginia Coal Mine Wars, 1920–1921." *West Virginia History* 50 (1991): 1–24.

"The Lebanon War: Operation Peace for Galilee (1982)." *Israeli Missions Around the World* (blog), n.d. http://embassies.gov.il/MFA/AboutIsrael/history/Pages/Operation %20Peace%20for%20Galilee%20-%201982.aspx.

Lee, Erika. *The Making of Asian America: A History*. New York: Simon & Schuster, 2015.

Lee, Robert G. *Orientals: Asian Americans in Popular Culture*. Asian American History and Culture. Philadelphia: Temple University Press, 1999.

Leeds, J. R. Memorandum to Chief of History and Research, DCNO (Air). "Aviation Personnel Fatalities in World War II," April 18, 1947. Naval History and Heritage Command Archives, Washington DC https://www.history.navy.mil/research/library/online-reading -room/title-list-alphabetically/a/aviation-personnel-fatalities-in-world-war-ii.html.

Leghorn, Richard S. "Aerial Reconnaissance." In *Selected Readings in Aerial Reconnaissance*, edited by Amrom Katz, 6–13. Santa Monica, CA: Rand Corporation, 1951.

Light, James, and David Welch. *Intelligence and the Cuban Missile Crisis*. London: Frank Cass, 1998.

Lindsey, Robert. "Al Ellis: Have Drone, Will Travel." *New York Times*, October 9, 1983. https://www.nytimes.com/1983/10/09/weekinreview/al-ellis-have-drone-will-travel .html.

Loeb, Vernon. "Alleged Beating of Prisoners Sparks Inquiry; Military Defends February 4 Missile Strike." *Washington Post*, February 12, 2002.

MacKenzie, Donald. *Inventing Accuracy: A Historical Sociology of Nuclear Missile Guidance*. Cambridge, MA: MIT Press, 1993.

MacKenzie, Donald, and Judy Wajcman. *The Social Shaping of Technology*. Buckingham, UK: Open University Press, 1999.

Macrakis, Kristie. "Technophilic Hubris and Espionage Styles during the Cold War." *Isis* 101, no. 2 (June 2010): 378–385.

Magnuson, Ed. "Israel Strikes at the PLO." *Time*, June 21, 1982. http://content.time.com /time/magazine/article/0,9171,925452,00.html.

Manjikian, Mary. "Becoming Unmanned: The Gendering of Lethal Autonomous Warfare Technology." *International Feminist Journal of Politics* 16, no. 1 (2013): 48–65.

Masco, Joseph. *Nuclear Borderlands: The Manhattan Project in Post–Cold War New Mexico*. Princeton, NJ: Princeton University Press, 2006.

———. *The Theater of Operations: National Security Affect from the Cold War to the War on Terror*. Durham, NC: Duke University Press, 2014.

Mbembe, Achille. "Necropolitics." Translated by Libby Meintjes. *Public Culture* 15, no. 1 (2003): 11–40.

McDougall, Walter. *The Heavens and the Earth: A Political History of the Space Age*. New York: Basic Books, 1984.

McLuhan, Marshall. *Understanding Media: The Extensions of Man* Cambridge, MA: MIT Press, 1994.

McNeill, William. *The Pursuit of Power: Technology, Armed Force, and Society since A.D. 1000*. Chicago: University of Chicago Press, 1982.

Miller, Andrea. "(Im)Material Terror: Incitement to Violence Discourse as Racializing Technology in the War on Terror." In *Life in the Age of Drone Warfare*, edited by Lisa Parks and Caren Kaplan, 112–133. Durham, NC: Duke University Press, 2017.

Miller, Roger. *Billy Mitchell: 'Stormy Petrel of the Air.'* Washington, DC: Office of Air Force History, 2004.

Mills, C. Wright. *The Power Elite*. New York: Oxford University Press, 1956.

Mitchell, W.J.T., and Mark B. N Hansen. *Critical Terms for Media Studies*. Chicago: University of Chicago Press, 2010.

MND-B Soldiers Kill Two Terrorists. Defense Visual Information Distribution Service, April 9, 2008. https://www.dvidshub.net/video/37449/mnd-b-soldiers-kill-two-terrorists#.U2-TGnavj54.

Moise, Edwin. *Tonkin Gulf and Escalation of the Vietnam War*. Chapel Hill: University of North Carolina Press, 1996.

Murray, Williamson. "Strategic Bombing: The British, American, and German Experiences." In *Military Innovation in the Interwar Period*, edited by Allen Millett and Williamson Murray, 96–143. Cambridge: Cambridge University Press, 1996.

Newcome, Laurence R. *Unmanned Aviation: A Brief History of Unmanned Aerial Vehicles*. Reston, VA: American Institute of Aeronautics and Astronautics, 2004.

Northrop-Grumman Media. "Targets Fact Sheet." Los Angeles: Northrop Grumman Corporation, 2014. http://www.northropgrumman.com/Capabilities/BQM74EAerialTarget/Documents/TGTS-Fact-Sheet.pdf.

Ohnuki-Tierney, Emiko. *Kamikaze, Cherry Blossoms, and Nationalisms: The Militarization of Aesthetics in Japanese History*. Chicago: University of Chicago Press, 2002.

———. *Kamikaze Diaries: Reflections of Japanese Student Soldiers*. Chicago: University of Chicago Press, 2007.

Oldenziel, Ruth. "Islands: The United States as Networked Empire." In *Entangled Geographies: Empire and Technopolitics in the Global Cold War*, edited by G. Hecht, 13–42. Cambridge, MA: MIT Press, 2011.

Packer, Jeremy, and Josh Reeves. "Romancing the Drone: Military Desire and Anthropophobia from SAGE to Swarm." *Canadian Journal of Communication* 38, no. 3 (September 14, 2013). https://doi.org/10.22230/cjc.2013v38n3a2681.

Parikka, Jussi. *Insect Media: An Archaeology of Animals and Technology*. Minneapolis: University of Minnesota Press, 2010.

Parks, Lisa. "Drones, Vertical Mediation, and the Targeted Class." *Feminist Studies* 42, no. 1 (2016): 227–235.

Parks, Lisa, and Caren Kaplan, eds. *Life in the Age of Drone Warfare*. Durham, NC: Duke University Press, 2017.

Parsch, Andreas. "MQM-105," 2002. *Directory of U.S. Military Rockets and Missiles*. http://www.designation-systems.net/dusrm/m-105.htm.

Pichirallo, Joe. "'Extraordinary' Process Used to Buy Navy Drone." *Washington Post*, July 28, 1988.

Purkiss, Jessica, Jack Serle, and Abigail Fielding-Smith. "US Counter Terror Air Strikes Double in Trump's First Year." *Bureau of Investigative Journalism*, December 19, 2017. https://www.thebureauinvestigates.com/stories/2017-12-19/counterrorism-strikes-double-trump-first-year.

"Reconnaissance: Cameras Aloft: No Secrets Below." *Time*, December 28, 1962. http://content.time.com/time/subscriber/article/0,33009,827960-1,00.html.

Ricks, Thomas E. "5 Big Problems with the Drone Programs." *Foreign Policy—Best Defense* (blog), December 10, 2015. https://foreignpolicy.com/2015/12/10/5-big-problems-with -the-drone-programs/.

Risen, James. "Threats and Responses: Hunt for Suspects; An American Was among 6 Killed by U.S., Yemenis Say." *New York Times*, November 8, 2002. https://www.nytimes.com /2002/11/08/world/threats-responses-drone-attack-american-was-among-6-killed-us -yemenis-say.html.

Rodrigues, Louis J. "Unmanned Aerial Vehicles: DOD's Acquisition Efforts." Testimony before U.S. House of Representatives, Committee on National Security. Washington, DC: Government Accountability Office, April 9, 1997. http://careersdocbox.com/US _Military/73538263-Gao-vehicles-unmanned-aerial-dod-s-acquisition-efforts.html.

Roland, Alex. *The Military-Industrial Complex*. Washington, DC: American Historical Association, 2001.

Rosen, Steven, ed. *Testing the Theory of the Military-Industrial Complex*. Lexington, MA: Lexington Books, 1973.

Rosenblueth, Arturo, Norbert Wiener, and Julian Bigelow. "Behavior, Purpose, and Teleology." *Philosophy of Science* 10, no. 1 (1943): 18–24.

Rowe, Peter. "His and Hers: Robert Schwanhausser's Life Has Two Big Chapters: One as a Man and Now One as a Woman." *San Diego Union Tribune*, June 17, 2007. http://legacy .sandiegouniontribune.com/uniontrib/20070617/news_mz1c17gender.html.

Sanders, Ralph. "An Israeli Military Innovation: UAVs." *Joint Force Quarterly* 33 (2002): 114–118.

Satia, Priya. "Drones: A History from the British Middle East." *Humanity* 5, no. 1 (Spring 2014): 1–31.

Scarborough, Rowan. "U.S. Strike Likely Killed Top Terrorist; Finance Official Thought Dead." *Washington Times*, February 13, 2002.

Schemmer, Benjamin. "Foreword." In *Lightning Bugs and Other Reconnaissance Drones: The Can-Do Story of Ryan's Unmanned Spy Plane*, by William Wagner. Fallbrook, CA: Aero Publishers; Armed Forces Journal International, 1982.

Schmitt, Carl. *The Nomos of the Earth: In the International Law of Jus Publicum Europaeum*. Translated by G. L. Ulmen. New York: Telos Press Publishing, 2006.

Schrader, Esther. "Response to Terror; Military to Probe Alleged Abuse of Afghan Prisoners; War: Rumsfeld Orders an Inquiry into Reports That U.S. Troops Beat Villagers Taken in Raid. Meanwhile, Officials Deny CIA Missile Strike Killed Civilians." *Los Angeles Times*, February 12, 2002.

Service Test in Field of TDR1—WWII, Torpedo Drone 30770. Periscope Film LLC, 2012. https://www.youtube.com/watch?v=8RQcUtzAe98.

Service Test of Assault Drone. Steven F. Udvar-Hazy Air and Space Center, Chantilly, VA, 1944.

Shanker, Thom, and James Risen. "A Nation Challenged: Raid's Aftermath; U.S. Troops Search for Clues to Victims of Missile Strike." *New York Times*, February 11, 2002. https://www.nytimes.com/2001/09/11/national/a-nation-challenged.html.

Shaw, Ian G. R. "Predator Empire: The Geopolitics of US Drone Warfare." *Geopolitics* 18, no. 3 (July 2013): 536–559.

Shaw, Ian G. R., and Majed Akhter. "The Unbearable Humanness of Drone Warfare in FATA, Pakistan." *Antipode* 44, no. 4 (September 2012): 1490–1509. https://doi.org/10 .1111/j.1467-8330.2011.00940.x.

Siegert, Bernhard. *Cultural Techniques: Grids, Filters, Doors, and Other Articulations of the Real*. Translated by Geoffrey Winthrop-Young. New York: Fordham University Press, 2015.

Simondon, Gilbert. *Du Mode d'Existence des Objets Techniques*. Paris: Aubier, 1958.

Singer, Peter. *Wired for War: The Robotics Revolution and Conflict in the Twenty-first Century.* New York: Penguin Press, 2010.

"Sinking the *Ostfriesland.*" *New York Times,* July 23, 1921. https://www.nytimes.com/1921/07 /23/archives/sinking-the-ostfriesland.html.

Sisk, Richard. "Mattis' Pet Peeve: Calling Drones 'Unmanned Aerial Vehicles.'" *Military. Com: Defense Tech* (blog), February 21, 2018. https://www.military.com/defensetech /2018/02/21/mattis-pet-peeve-calling-drones-unmanned-aerial-vehicles.html.

Smith, Merritt Roe, ed. *Military Enterprise and Technological Change: An American Perspective.* Cambridge, MA: MIT Press, 1985.

Solnit, Rebecca. *Savage Dreams: A Journey into the Landscape Wars of the American West.* Berkeley: University of California Press, 2014.

Spark, Nick T. "Command Break: The Battle Over America's Secret WWII Cruise Missile." *Proceedings* (United States Naval Institute) (2005). Accessed September 2, 2018. http:// stagone.org/?page_id=20.

Struck, Doug. "Casualties of U.S. Miscalculations; Afghan Victims of CIA Missile Strike Described as Peasants, Not Al Qaeda." *Washington Post,* February 11, 2002.

Suchman, Lucy. *Human-Machine Reconfigurations: Plans and Situated Actions.* Cambridge: Cambridge University Press, 2007.

Suchman, Lucy, Karolina Follis, and Jutta Weber. "Tracking and Targeting: Sociotechnologies of (In)Security." *Science, Technology, and Human Values* 42, no. 6 (November 2017): 983–1002. https://doi.org/10.1177/0162243917731524.

Swan, Bobbi. "Whatever Happened to Bobbi Swan?" *TG Forum* (blog), May 23, 2009. https://tgforum.com/wordpress/whatever-happened-to-bobbi-swan.

Szasz, Ferenc Morton. *The Day the Sun Rose Twice: The Story of the Trinity Site Nuclear Explosion, July 16, 1945.* Albuquerque: University of New Mexico Press, 1984.

Tanaka, Toshiyiki, and Marilyn Blatt Young. *Bombing Civilians: A Twentieth-Century History.* New York: New Press, 2009.

Taubman, Philip. *Secret Empire: Eisenhower, the CIA, and the Hidden Story of America's Space Espionage.* New York: Simon & Schuster, 2003.

Tchen, John Kuo Wei, and Dylan Yeats, eds. *Yellow Peril! An Archive of Anti-Asian Fear.* London: Verso Books, 2014.

"TDR-1 EDNA III." *National Naval Aviation Museum* (blog), n.d. http://www .navalaviationmuseum.org/attractions/aircraft-exhibits/item/?item=tdr.

Tesla, Nikola. "Tesla's New Device like Bolts of Thor." *New York Times,* December 8, 1915. http://www.teslacollection.com/tesla_articles/1915/new_york_times/nikola_tesla/tesla _s_new_device_like_bolts_of_thor.

Thompson, Charis. *Making Parents: The Ontological Choreography of Reproductive Technologies.* Cambridge, MA: MIT Press, 2005.

Thompson, Coleman. "Fleet Composite Squadron 6 Deactivates." *Naval Station Norfolk News* (blog), August 8, 2008. http://www.navy.mil/submit/display.asp?story_id=38993.

"U-2S/TU-2S." *U.S. Air Force Fact Sheets* (blog), September 25, 2015. http://www.af.mil /About-Us/Fact-Sheets/Display/Article/104560/u-2stu-2s/.

United Nations General Assembly. "The Situation in the Middle East." December 16, 1982. https://www.un.org/documents/ga/res/37/a37r123.htm.

Uricchio, William. "Television's First Seventy-five Years: The Interpretative Flexibility of a Medium in Transition." In *The Oxford Handbook of Film and Media Studies,* edited by Robert Kolker, 286–305. Oxford: Oxford University Press, 2008.

U.S. Congress. Authorization for Use of Military Force, Pub. L. No. 107–40, 115 STAT. 224 (2001). https://www.congress.gov/107/plaws/publ40/PLAW-107publ40.pdf.

Virilio, Paul. *War and Cinema: The Logistics of Perception.* Translated by Patrick Camiller. 1989. Reprint. London: Verso Books, 2009.

Wagner, William. *Lightning Bugs and Other Reconnaissance Drones: The Can-Do Story of Ryan's Unmanned Spy Plane.* Fallbrook, CA: Aero Publishers; Armed Forces Journal International, 1982.

Wagner, William, and William P. Sloan. *Fireflies and Other UAVs (Unmanned Aerial Vehicles).* Arlington, TX: Aerofax, 1992.

Weber, Samuel. *Targets of Opportunity: On the Militarization of Thinking.* New York: Fordham University Press, 2005.

Weinberger, Caspar W. *Fighting for Peace: Seven Critical Years in the Pentagon.* New York: Warner Books, 1990.

Werrell, Kenneth. *The Evolution of the Cruise Missile.* Maxwell Air Force Base, AL: Air University Press, 1985.

Whittle, Richard. *Predator: The Secret Origins of the Drone Revolution.* New York: Henry Holt and Company, 2014.

———. "Predator's Big Safari." Mitchell Institute for Airpower Studies, August 2011. http://secure.afa.org/Mitchell/reports/MP7_Predator_0811.pdf.

Wicker, Tom. "M'Namara Insists Offensive Arms Are Out of Cuba." *New York Times,* February 7, 1963. https://www.nytimes.com/1963/02/07/archives/mnamara-insists-offensive -arms-are-out-of-cuba-declares-on-tv-that.html.

Wilcox, Lauren. "Embodying Algorithmic War: Gender, Race, and the Posthuman in Drone Warfare." *Security Dialogue* 48, no. 1 (February 2017): 11–28.

Winner, Langdon. "Upon Opening the Black Box and Finding It Empty: Social Constructivism and the Philosophy of Technology." *Science, Technology, & Human Values* 18, no. 3 (1993): 362–378.

Woods, Chris. "The Story of America's Very First Drone Strike." *The Atlantic,* May 30, 2015. https://www.theatlantic.com/international/archive/2015/05/america-first-drone-strike -afghanistan/394463/.

Woods, Chris, and Christina Lamb. "CIA Tactics in Pakistan Include Targeting Rescuers and Funerals." *Bureau of Investigative Journalism,* February 4, 2012. https://www .thebureauinvestigates.com/stories/2012-02-04/cia-tactics-in-pakistan-include-targeting -rescuers-and-funerals.

Zaloga, Steve. *Unmanned Aerial Vehicles: Robotic Air Warfare, 1917–2007.* New Vanguard 144. Oxford: Osprey Publishing, 2008.

Index

Page numbers in *italics* refer to photographs.

About the Author

KATHERINE CHANDLER is an assistant professor in the Culture and Politics Program at the Walsh School of Foreign Service, Georgetown University. She received her PhD in rhetoric from the University of California, Berkeley.